北京古代
建筑精粹

GEMS OF BEIJING ANCIENT
ARCHITECTURE

# 北京古代建筑精粹

GEMS OF BEIJING ANCIENT ARCHITECTURE VOLUME II

北京市文物局
《北京古代建筑精粹》编委会 编

北京出版社 出版集团
BEIJING PUBLISHING HOUSE (GROUP)
北京美术摄影出版社

# 下 卷 目 录

| | |
|---|---|
| 寺观 | 8 |
| 北京的寺观 | 10 |
| 雍和宫 | 16 |
| 潭柘寺 | 30 |
| 戒台寺 | 40 |
| 云居寺塔及石经 | 46 |
| 碧云寺 | 54 |
| 万寿寺 | 62 |
| 十方普觉寺(卧佛寺) | 68 |
| 妙应寺白塔 | 74 |
| 法源寺 | 78 |
| 大觉寺 | 82 |
| 大慧寺 | 90 |
| 广化寺 | 96 |
| 广济寺 | 100 |
| 觉生寺(大钟寺) | 104 |
| 法海寺 | 108 |
| 灵岳寺 | 114 |
| 承恩寺 | 118 |
| 普度寺 | 120 |
| 智化寺 | 122 |
| 天宁寺塔 | 128 |
| 真觉寺金刚宝座塔 (五塔寺塔) | 132 |
| 银山塔林 | 136 |
| 万佛堂、孔水洞石刻及塔 | 138 |
| 清净化城塔 | 142 |
| 慈寿寺塔 | 146 |
| 燃灯塔 | 148 |
| 姚广孝墓塔 | 150 |
| 镇岗塔 | 151 |
| 良乡塔 | 151 |
| 白云观 | 152 |
| 东岳庙 | 160 |
| 大高玄殿 | 166 |
| 火德真君庙 | 170 |
| 关岳庙 | 172 |
| 牛街礼拜寺 | 174 |
| 东四清真寺 | 182 |
| | |
| 城垣 | 184 |
| 北京的城垣 | 186 |
| 万里长城——八达岭 | 192 |
| 长城——司马台段 | 200 |
| 居庸关　云台 | 210 |
| 正阳门 | 220 |
| 北京城东南角楼 | 224 |
| 德胜门箭楼 | 225 |
| | |
| 陵墓 | 226 |
| 北京的陵墓 | 228 |
| 十三陵 | 234 |
| 景泰陵 | 266 |
| 醇亲王墓 | 268 |
| | |
| 其他 | 272 |
| 北京的其他古代建筑 | 274 |
| 卢沟桥 | 278 |
| 永通桥及石道碑 | 284 |
| 琉璃河大桥 | 286 |
| 万宁桥 | 287 |
| 朝宗桥 | 288 |
| 湖广会馆 | 290 |
| 安徽会馆 | 294 |
| 正乙祠 | 296 |
| 北京鼓楼、钟楼 | 298 |
| 燕墩 | 304 |
| 健锐营演武厅 | 306 |
| 爨底下村古代建筑群 | 310 |
| | |
| 附录 | 314 |
| 北京市各级别文物保护单位名录 | 314 |
| 北京地区古代建筑修缮工程大事记 | 337 |
| 后记 | 341 |

# Contents of Volume II

MONASTERIES AND TEMPLES 8
Monasteries and Temples in Beijing 13
Yonghegong Lama Temple 16
The Pool and Cudrania Temple 30
Jietai Temple 40
Yunju Temple 46
The Temple of Azure Clouds 54
Wanshou (Long Life) Temple 62
The Temple of Universal Awakening 68
The White Stupa in the Miaoying Temple 74
Fayuan Temple 78
Dajue Temple 82
Dahui Temple 90
Guanghua Temple 96
Guangji Temple 100
Juesheng Temple 104
Fahai Temple 108
Lingyue Temple 114
Cheng'en Temple 118
Pudu Temple 120
Zhihua Temple 122
The Pagoda at the Temple
of Celestial Tranquility 128
The Diamond Throne Pagoda at the
Temple of True Awakening 132
Silver Mountain Pagoda Forest 136
Wanfotang and Kongshui Cave 138
Qingjing Huacheng Pagoda 142
The Pagoda at the Temple of
Benevolence and Longevity 146
Randeng Pagoda 148
The Stupa for Yao Guangxiao 150
Zhengang Pagoda 151
Liangxiang Pagoda 151
The White Cloud Daoist Temple 152
Dongyue Temple 160
Dagaoxuandian 166
The Temple of the Perfect Sovereign
of the Virtue of Fire 170
Guanyue Temple 172

Niujie (Ox Street) Mosque 174
Dongsi Mosque 182

CITY DEFENSE INSTALLATIONS 184
City Defense Installations in Beijing 189
The Great Wall in Badaling Section 192
The Great Wall in Simatai Section 200
Juyongguan and Cloud Terrace 210
Zhengyangmen 220
The Southeastern Corner Tower 224
Deshengmen Embrasure Watchtower 225

TOMBS 226
Tombs in Beijing 231
The Ming Tombs 234
Jingtailing 266
Prince Chun's Tomb 268

THE OTHERS 272
Other Ancient Architecture in Beijing 276
Lugou Bridge 278
Yongtong Bridge 284
Liulihe Bridge 286
Wanning Bridge 287
Chaozong Bridge 288
Huguang Guild Hall 290
An'hui Guild Hall 294
Zhengyi Ancestral Temple 296
Drum Tower and Bell Tower of Beijing 298
Yan Pier 304
Ancient Qing Dynasty Fortress 306
The Ancient Village Chuandixia 310

Appendices 314
A List of Protected Residences of
Various Levels of Beijing 314
Chronicle of Events Concerning the
Renovation Works of Ancient Buildings
in Beijing Area 337
Postscript 341

寺观

# MONASTERIES
# AND TEMPLES

# 北京的寺观

宗教是我国古代社会生活的重要组成部分，宗教建筑——寺观一方面是神灵的圣地，同时也是人们的公共场所。作为皇都，北京现存古代建筑中有大量的佛教、道教、伊斯兰教寺观。这些宗教建筑有一个共同的特点是保存了大量皇家敕建的寺院道观，是其他任何一个地方都无法比拟的。另外一个共同点就是主要为明清以来的建筑，北京现存的宗教建筑有几百座之多，多数为明、清两代修建的(只有个别的佛塔、经幢、佛殿是较早期的作品)，尤其清代康乾时期对前代寺庙多有修、改、扩建等工程。因此，在建筑的总体布局以及个体的做法上，都体现了明、清时期佛教建筑的特征。这些都是因为北京是封建王朝后期都城。

北京的宗教建筑由于教义和使用要求不同，总体布局和建筑式样也各有特色。

## 一、佛教建筑

中国佛教主要分为三大派，主要是汉传佛教(大乘佛教)、藏传佛教(密宗)、南传佛教(上座部)，而北京地区主要以汉传佛教和藏传佛教为主，所以佛教建筑也主要是汉传佛教和藏传佛教建筑，包括佛寺、佛塔和石造像等。

### 1.汉传佛教建筑

汉传佛教历史悠久，在中国流传最广，北京的宗教建筑也以佛寺为最多。据古籍中记载，北京有史可考的最古老的寺院是西山的潭柘寺，始建于晋代永嘉元年(307年)，原名嘉福寺，所以流传有"先有潭柘，后有幽州"的谚语。唐代早期，北京所处的幽州已经成为当时全国重要的佛教兴盛地，隋唐时期又出现了云居寺、戒台寺、法源寺等著名寺庙。辽统治者十分重视宗教的作用，尤其是十分支持佛教的发展，甚至统治者本身也十分虔诚。在此期间，传统名寺继续香火鼎盛、发展壮大，留下了一些流传至今的著名的塔、幢等建筑，如天宁寺、银山塔林等建筑。元明清时期，统治者大力提倡宗教信仰，以致在北京出现了大

量的大大小小的宗教建筑，尤其是内城甚至每几条胡同就有一座公共的宗教建筑。根据《乾隆京城全图》统计，城墙以内的大大小小的寺庙就有1000多座，并且这一时期修建了很多皇家敕建的宗教建筑，如法海寺、承恩寺、碧云寺、卧佛寺和万寿寺等。这些庙宇不但等级高而且规模大、建筑精，在北京宗教史上达到了高峰。另外，清政府利用宗教统治蒙藏地区，也确实达到了安边定国的目标。

明、清佛寺的布局一般是由中轴线上的门、殿和两侧的配殿、廊庑组成严格对称的多进院落。主轴线上最前方为山门，门内左右有钟、鼓楼，门北正面为天王殿，多做成穿堂。天王殿北为正殿，一般称大雄宝殿，是寺中最主要的建筑。正殿左右多有配殿，配殿有时可做成二层楼房。正殿后视佛寺规模还可以有第二层或第三层殿，最后常常以二层的藏经楼结束。殿东西侧除配殿外还有廊庑，围成殿庭。廊庑可做斋堂、僧房等用途。大型佛寺往往在中轴线建筑群两侧建若干小院，做僧房、方丈院等，形成左中右三条轴线，一般称为中路、东路、西路。北京大型佛寺如潭柘寺、碧云寺、卧佛寺都是有三路院落的大寺院。小型的寺庙，一般只有一到两进院落，一进山门迎面就是大殿，两厢为配殿，二进的后面还有一座后殿。

在佛寺中还常常布置一些附属建筑，如山门外加牌坊、影壁、石狮、旗杆、经幢、碑等。寺内设钟、磬、木鱼、云板、铜炉、碑碣和其他石刻及铜器，以增加寺院的宗教气氛。殿内除佛像外，往往在四壁绘有壁画，并以佛坛为中心，悬挂帷幔，布置香案、供具、幡幢等。低垂的帷幔使殿内光线幽暗，对形成佛殿内神秘、肃穆的气氛是不可缺少的。有些寺庙还有精美的园林。

元大都时城内就建有许多巨大的佛寺，可以确认的是建于至元七年(1270年)的柏林寺。现在该寺南北长约150米，恰是元大都两条胡同间的标准距离，东西宽为长的3/2，符合当时一般佛寺平面比例。该寺毁于元末，明正统时重建，清康熙、乾隆时重修。明代佛寺中保存较完整的是智化寺，此寺为正统时太监王振所造，虽然赐名为寺，实际是王振的家祠。此寺建筑全用黑琉璃瓦屋顶，明

代尊崇玄武，色以黑为贵，故明代大寺多用黑瓦。寺中之智化殿和万佛阁的藻井华贵精美，在民国年间被盗卖至美国。另外，明代建于景泰三年(1452年)的隆福寺，占地约26亩。该寺还是京城东西南北城四大庙会中的主体。1910年大火烧毁不少殿宇，20世纪50年代后陆续拆除，改建商场。主殿正觉殿内精美的木雕藻井现在北京市古代建筑博物馆内展出。

## 2.藏传佛教寺院

喇嘛教即藏传佛教，是我国蒙、藏两个兄弟民族信奉的宗教，所以喇嘛庙的建筑也因宗教内容的特点和民族的特点而和佛寺有所不同。北京的喇嘛教建筑是从元代开始出现的。清朝为抚绥蒙、藏，以抬高藏传佛教的地位为国策，从而大量兴建喇嘛庙。喇嘛教建筑一般有两种形式，一种是和佛寺相近的宫室式木建筑；另一种是属于碉房式的砖石建筑。北京的雍和宫和东、西黄寺都属于前者，只有颐和园后山的一组喇嘛寺是碉房式的。木建筑的喇嘛寺仍旧采用了四合院式布局，寺庙前半部的山门、天王殿、大殿都和佛寺差不多。但大殿以后的部分常有高大而雄伟的建筑，在布局上也有所变化，例如雍和宫的后部采用三殿并列的制度，用复道将高大的万福阁和两侧的永康阁、延宁阁连接起来，气势十分壮丽。喇嘛教的塔不同于佛塔，它是由印度的窣屠坡演化而来的。中国的喇嘛塔也有作为僧侣坟墓的，但建筑在寺庙中的喇嘛塔，却和佛塔的作用相同。

清代新建了不少藏传佛教寺院，目前还有两座等级最高，保存最完整，最具有藏传佛教特色的寺庙，一座是故宫内的中正殿和雨花阁，另一座是雍和宫。雨花阁建在皇宫内，金瓦屋面，高耸于后宫之中，表示清代皇帝对本教的推崇。其造型融合了汉、藏形式，是一座教义与艺术完美结合的建筑。雍和宫原是雍正未称帝前的雍亲王府，雍正三年(1725年)改名雍和宫，乾隆九年(1744年)改建成京城最大的喇嘛庙。其中颇具特色的建筑有法轮殿(空间脱胎于藏式殿，屋顶起五座天窗，象征五方佛)、班禅楼和戒台楼(外观为二层楼阁，内部空间为蒙、藏式大殿)。此外，雍和宫还是一座正规的宗教大学，按藏传佛教的修习制度，设有经义、仪轨密宗、医药、历算四学殿。

## 3.塔

塔原称"窣屠坡"，是随佛教从印度传入中国的。建塔之初，是为了纪念佛教创始人释迦牟尼，尔后，成了历代高僧圆寂后埋藏、保存其"舍利子"("舍利"为梵文的译音，其含义即身骨，在佛教中舍利是一种至高无上的神圣物品)的建筑。随着佛教在中国的传播，佛塔也随之在各地建造并普及，逐渐成为中国宗教的代表性建筑。

在北京众多的古代建筑中，塔的建造数量和建筑规模都占有重要的地位，据不完全统计，北京现存的塔约有近300座。北京建塔的年代，从史料记载推断大体上始于隋朝的弘业寺(今北京天宁寺的前身)，现存年代最早的则是唐塔(房山区云居寺现存8座唐代石塔)。辽金时期是北京佛塔发展的重要阶段，代表作有北京天宁寺塔、昌平银山塔林的五座大型塔群和北京良乡昊天塔等。元代北京盛行喇嘛教，最具典型的喇嘛塔当属妙应寺白塔。明朝北京的佛塔在建造数量和建筑质量以及种类上都达到了新的飞跃，其代表作有慈寿寺塔、真觉寺金刚宝座塔等。清代，佛教开始走向衰落，建塔的数量较前朝要少得多，代表作有静宜园内的宗镜大昭庙琉璃塔、清净化城塔、北海永安寺白塔等。

北京古塔的形制种类很多，主要建筑形式大致可分为密檐式、楼阁式、覆钵式和金刚宝座式等。材质以砖塔、石塔为主，砖木混合塔、琉璃塔次之，其他材质的古塔较少。

## 二、道教建筑

道教是我国古代自创的宗教，唐、宋时由于皇帝提倡，曾经大修道观，兴盛一时，以后各代也都与佛教并行，没有偏废。但道观建筑并没有什么突出的特点，基本上和佛寺相似，多是模拟佛寺的殿堂建造成的，所不同之处是没有塔、幢等佛教特有的建筑物而已。

北京的道教建筑，根据文献记载，远在唐代就有了,如北京火德真君庙，即火神庙，始建于唐贞观年间，但真正大规模地修建道观是从元代开始的，元大都时最大的道观是金中都的长春宫，元末被毁，明清在其东侧重建为白云观。道教在明代永乐和嘉靖时期一度受到皇帝重视，修建了几处大型道观，其中很重要的一座是建于明成化十七年(1481年)的大慈延福宫，又名三官庙，现在也只剩后面的一座殿堂和少数配房。这三座殿堂为歇山式黑色琉璃顶，内部有木雕藻井，据此可见当年道观的盛况。清代道观位于钟楼后的宏恩观，规模较大，还有较多遗存。

## 三、伊斯兰教建筑

伊斯兰教是我国回族、维吾尔族、哈萨克族等许多少数民族信仰的宗教，从唐代开始传入中国。全国各地建有很多伊斯兰教的寺院，称为清真寺或礼拜寺。

唐辽之际已有阿拉伯人进入北京地区，元代以后入住大都者更多，他们融入中华民族形成了回族，族教一体，皇都中也建有一些清真礼拜寺。北京的伊斯兰教寺庙现存共有80余座，最早的伊斯兰教礼拜寺是宣武门外的牛街礼拜寺，有记载始建于辽代。元大都城内住有很多阿拉伯色目人，也应当有相当数量的礼拜寺。现存东四清真寺一说是始建于元至正六年(1346年)，另一说始建于明景泰元年(1450年)，推测是元时有寺，但是阿拉伯式，明初被毁，景泰时重建为传统木构建筑。

北京地区现存的明初以前的清真寺多做穹隆顶、尖拱门窗，带有浓厚的中亚伊斯兰建筑风格，明中期以后逐渐和汉族传统木构建筑形式结合。一般清真寺建筑有礼拜殿、望月楼、邦克楼、浴室、教室等。礼拜殿是主要建筑，布置在中轴线的中部，由前廊、拜殿和后窑殿三部分组成，外观是汉族木构勾连搭殿堂的形式。只有窑殿内设有礼拜龛，有很强的阿拉伯建筑风格。望月楼是供每年斋月开始前一天和最后一天由阿訇登楼望月以决定开斋和封斋时辰用的，邦克楼又叫唤醒楼，是供每天日出、中午和日落时阿訇登楼呼唤教徒前来做礼拜用的，它们一般也都设在中轴线上，其他建筑对称地布置在两侧。中国伊斯兰教清真寺很大的一个特点是因为窑殿内的龛背向西方，信徒礼拜时能朝向麦加，所以拜殿要面向东方。清真寺建筑不用动物形象作装饰题材，纹样多为几何图案或阿拉伯文字组成，虽然也采用汉族传统的沥粉贴金彩画和各种木雕，还是有其特殊的装饰风格的。

# Monasteries and Temples in Beijing

The existing ancient buildings in Beijing include a large number of Buddhist, Taoist, Islamist and Lamaist temples. The large number of temples built under imperial order can be found nowhere else except in Beijing and most of them were built in the Ming and Qing Dynasties. Now in Beijing there still exist hundreds of religious buildings, among which the earliest built Buddhist halls so far known date back only to the Yuan Dynasty, such as the Halls of Lingyue Temple and Lingyan Temple. Most others were built in the Ming and Qing Dynasties except several stupas, sutra libraries and Buddhist halls built earlier. The religious buildings in Beijing differ from one another in overall layout and architectural style as they served different doctrines and were used for different purposes.

## 1. Buddhist Architecture

Buddhism in China mainly consists of three schools: Han-Chinese Buddhism (Mahayana), Tibetan Buddhism (Tantrism) and Southern Buddern (Theravada). In Beijing Han-Chinese Buddhism and Tibetan Buddhism are the mainstream, so most of the local Buddhist buildings, such as temples, stupas and stone sculptures, are typical of the two Buddhist schools.

### (1) Han-Chinese Buddhist buildings

In Beijing the oldest temple recorded in ancient books is the Pool and Cudrania Temple in the Western Mountain. Formerly known as Jiafusi, it was first built in the 1st year of the Yongjia reign in the Jin Dynasty (307 A.D.), so it is often said that the Pool and Cudrania Temple appeared before the existence of Youzhou. Youzhou, where Beijing was located, had become the most important Buddhist holy land of China in the early Tang Dynasty and during the Sui and Tang Dynasties, many famous temples were built, such as the Yunju Temple, the Jietai Temple and the Fayuan Temple. Rulers of the Liao Dynasty attached great importance to the role of religions, especially of Buddhism, to which even themselves were pious. During this period, traditionally famous temples remained prosperous and expanding, and until now some well-known towers and houses are still in existence, such as Tianning Temple and Yinshan Pagodas. During the Yuan, Ming and Qing Dynasties, in response to rulers' calling for extensive belief in religion, religious buildings appeared one after another in Beijing, especially in the inner city where almost every couple of Hutongs share a public religious building. According to the statistics based on *The Beijing Map of the Qianlong Reign*, all kinds of temples within the city walls total more than 1,000. This period also witnessed the emergence of many Buddhist buildings built under imperial order, such as the Fahai Temple, the Cheng'en Temple, the Temple of Azure Clouds, the Temple of Universal Awakening, the Wangshou Temple etc, These temples, characterized by high rank, large scale and architectural excellence, mark a climax in the religious history of Beijing. In fact, the religion did help the Qing regime strengthen its governance of Mongolian and Tibetan regions and stabilize the country.

Most temples in the Ming and Qing Dynasties are multi-access courtyards, which include a primary gate and a primary hall on the middle axis, and wing halls, corridors and houses symmetrical on each side of the axis. In the very front of the axis is a front gate, in which are bell and drum towers on each side. Straight to the north of the gate is the Hall of Heavenly Kings, to the north of which is the main hall. As the main hall of the temple, it is usually called the Mahavira Hall, flanked by wing halls on the eastern and western sides, which may be two-storied buildings. Behind the main hall may be the second or third halls, depending on the size of the temple. The final part is often a two-storied Sutra Hall. In the east and west of the main hall are not only wing halls but also corridors and houses, all of which form a courtyard around the main hall. The corridors and houses can be used for dining hall and monks' dormitory. Large Buddhist temples often have many small courtyards on both sides of the central building complex, which are used to accommodate monks' dormitory and abbot's room. All the buildings form three axes respectively in the middle, the eastern and the western. In Beijing large temples that consist of such three axes include the Pool and Cudrania Temple, the Temple of Azure Clouds and the Temple of Universal Awakening. Small temples usually have only one or two courtyards one behind another. Just behind the front gate is the main hall, on each side of which are monks' dormitories. Buddhist temples are often dotted with some auxiliary buildings, such as archway, screen wall, stone lion, banner pole, pagoda, stone pillar and stele outside the temple.

## (2) Tibetan Buddhist buildings

In an attempt to pacify Mongolia and Tibet, the Qing regime adopted the national policy of upholding Tibetan Buddhism, which resulted in the mushrooming of Lamaseries. Lamaist architecture usually takes two forms: one is palace-style wooden building similar to Buddhist temple, the other is blockhouse-style stone chambers. The former is represented by the Yonghegong Lama Temple, the Donghuang Temple and the Xihuang Temple; the latter by the lamaseries in the back mountain of the Summer Palace. Wooden lamaseries are designed just as courtyards are and their front gate, Heavenly Kings Hall and main hall in the front are all similar to those in Buddhist temples, but behind the main hall often stand stately buildings in a different layout. Take the Yonghegong Lama Temple as an example. In its back part stand side by side three halls: the grand Wanfu Tower in the middle and the Yongkang Tower and the Yanning Tower each in the side. Connected by compound paths, the three halls look splendid. Many Tibetan Buddhist temples were built in the Qing Dynasty.

Among them the highest-ranking and best-preserved two most typical of Tibetan Buddhism are still in existence. One is the Hall of Complete Harmony and Rain Flower Tower in the Palace Museum, the other is the Yonghegong Lama Temple.

## (3) Pagodas

In Beijing, Pagodas are distinguished from other ancient buildings by their number and scale. According to incomplete statistics, now there are nearly 300 pagodas in Beijing. The first pagoda in Beijing, as implied in historical records, was built in the Hongye Temple (now known as Tianningsi) during the Sui Dynasty. The earliest built pagoda still in existence is Tang Pagoda (now there are 8 stone pagodas of the Tang Dynasty in the Yunju Temple, Fangshan District). The Liao and Jin periods witnessed a robust development of Buddhist pagodas in Beijing, of which the representatives include the Pagoda at the Temple of Celestial Tranquility, the five large groups of pagodas in Silver Mountain, Changping and the Haotian Pagoda in Lianxiang. By the Yuan Dynasty, Lamaism began to prevail in Beijing and the White Pagoda in Miaoying Temple built during that period was the most typical Lamaist pagoda. By the Ming Dynasty, Buddhist pagodas in Beijing reached a new high in number, quality and type, as evidenced by the pagoda of Cishou Temple and the Diamond Throne Pagoda at the Temple of True Awakening. Representative pagodas in the Qing Dynasty include the glazed pagoda in Zhongjing Dazhao Temple of Jingyi Garden, the Diamond Throne Pagoda of Jinghuacheng and the white pagoda in Yongan Temple of Beihai.

Ancient pagodas in Beijing were built in many different forms and types. Their architectural forms can be basically divided into dense-eaves type, pavilion type, inverted-bowl type and throne type. Most of them were made of bricks and stones, some were made of mixed bricks and wood or glazed tiles and few were made of other materials.

## 2. Taoist Architecture

The Taoist buildings, according to historical literature, date back to the Tang Dynasty. The Temple of the Perfect Sovereign of the Virtue of Fire is an example. Large-scale Taoist temples did not appear until the Yuan Dynasty. The largest Taoist temple in Yuan Dadu is Changchun Palace in Jinzhongdu, which was destroyed in the late Yuan Dynasty. By the Ming and Qing Dynasties another temple, called White Cloud Temple, was built in the east of the former site. Taoism was rendered great importance by emperors during the Yongle and Jiajing reigns of the Ming Dynasty when several large Taoist temples were built. An important one of them is Daciyanfu Palace, also known as Sanguan Temple, which was built in the 17th year (1481A.D.) of the Chenghua reign, but by now only the hall in the rear and a few skirt buildings are still in existence. The Hong'en temples, which were built behind the bell tower in the Qing Dynasty are of large scale and many buildings are still in existence now.

## 3. Islamic Architecture

As early as in the Tang and Liao Dynasties, Arabs had begun to enter Beijing. After the Yuan Dynasty, more Arabs came to live in the capital city. They integrated themselves and their religion into the Hui nationality of the Chinese nation. As a result, in the capital city appeared some mosques. Now there are more than 80 mosques in Beijing. The earliest built mosque lies in Niujie Street, Xuanwumenwai and according to historical records, it was built in the Liao Dynasty. In Yuan Dadu lived many Arabian Semu people and stood quite a lot of mosques. The existing Dongsi Mosque is said to be built either in the 6th year of the Zhizheng reign during the Yuan Dynasty (1346 A.D.) or in the 1st year of the Jingtai reign during the Ming Dynasty (1450 A.D.). Presumably, the mosque appeared in the Yuan Dynasty and was destroyed in the early Ming Dynasty. The newly built mosque during the Jingtai reign was traditionally wood-structured.

# 雍和宫
# Yonghegong Lama Temple

雍和宫位于北京市东城区雍和宫大街12号，是北京地区规模最大、保存最完好的藏传佛教寺院。1961年被公布为全国重点文物保护单位。

雍和宫创建于清康熙三十三年(1694年)，初为雍正帝胤禛即位前的府邸，胤禛即位后，作为潜邸，将其中的一半改为黄教上院，一半作为皇帝行宫。乾隆九年(1744年)，雍和宫改为正式的喇嘛庙，成为清政府管理全国喇嘛教事务的中心。建国后，对雍和宫进行了全面修缮，1981年作为宗教活动的场所正式对外开放，并成为著名的游览胜地。

雍和宫坐北朝南，占地约6.6万平方米，分为中、东、西三路。中路是主体建筑所在，前方为三座牌楼，经辇道，直通雍和宫正门昭泰门，之后为中路的五进院落，分别是天王殿、雍和宫、永佑殿(神御殿)、法轮殿和万佛阁。雍和宫东路有东书院、平安居、如意室、太和斋以及海棠院、花园等建筑。西路原为观音殿和关帝庙，此外还有六座阿嘉仓，位于昭泰门外辇道两侧，是历代活佛的佛仓(即住所)。

因其由王府改建而成，所以雍和宫整个寺院前后风格各异：前半部建筑疏朗开阔，是宫殿形状；后半部是庙宇风格，建筑紧凑有序，殿阁交错，并具有汉、蒙、满、藏诸民族建筑风格融为一体的特色。整个雍和宫布局严整，建筑壮观，建筑结构和造型艺术都有很高的水平，而且还有很多珍贵文物堪称绝世仅有。

The Yonghegong Lama Temple is located at No.12 Yonghegong Street in Dongcheng District. It is the largest and best-preserved Tibetan Buddhist Lama Temple in Beijing and was listed as a national key relic under special preservation in 1961.

It was first built in the 33rd year of Emperor Kangxi of the Qing Dynasty (1694 A.D.), as the residence of the Qing Dynasty Prince Yong. After he ascended the throne, it was converted to the lama temple of the Yellow Hat Sect of Lamaism and the palace for emperor's temporary dwelling. In the 9th year of the reign of Emperor Qianlong (1744 A.D.), it became a lamasery. After the People's Republic of China was founded, the Lama Temple was renovated completely. In 1981, it was opened to the public as the place for religion and became famous tourist attraction.

Facing south the temple occupies 66,000 square meters of land. It is divided into the middle, the eastern and the western axes. On the middle axis, from the front to the rear, are three exquisite archways, the Zhaotai Gate, the Hall of Heavenly Kings, the Yonghegong Hall, the Yongyou Hall, the Falun Hall and the Wanfo Hall.

雍和宫全貌
A Panorama of the Yonghegong Lama Temple

牌楼
The Archway

天王殿
The Hall of Heavenly Kings

原为王府时期正门，面阔五间，单檐歇山顶。殿内供奉弥勒佛、四大
天王及韦驮天王像。院内有碑亭两座，以及钟鼓楼。

御碑亭
The Imperial Tablet Pavilion

御碑亭内有著名的汉、满、蒙、藏四体文《喇嘛说》碑，是乾隆五十七
年(1792年)设立的，碑文为乾隆皇帝所撰，记述了喇嘛教的来源和清
朝政府对喇嘛教的政策。

雍和宫原为雍亲殿和中心，面阔七间，单檐歇山式，匾额"雍和宫"是用汉、满、蒙、藏四种文字写成，殿内供奉三世佛(药师佛、如来佛和阿弥陀佛)、观世音菩萨和弥勒佛以及十八罗汉。东西配楼、配殿四座，又称"四学殿"，即讲经殿、密宗殿、数学殿、药师殿，是僧人们学习显、密经典和数学、历学、医术的地方。院内还有一尊铸造于乾隆十二年(1747年)的铜鼎，清朝皇帝每次到雍和宫拜佛，都是用它进香。铜鼎后面是御碑亭和明代万历年间铜制须弥山。

雍和宫及大殿前的须弥山
The Mount Sumeru in Front of the
Yonghegong Hall

永佑殿(神御殿)
The Yongyou Hall

永佑殿佛像
Buddhist Statues in the Yongyou Hall

殿内正中供奉着三尊佛像，正中为无量寿佛、左为药师佛、右为狮吼
佛，佛像高2.35米，檀香木雕刻而成。殿西墙挂着绿度母补绣像，据
说是乾隆之母孝圣皇太后亲手补绣。东墙悬挂着白度母画像。

法轮殿殿顶天窗及塔式宝顶
Windows on Top of the Falun Hall and Pagodas

法轮殿是雍亲王府时期的正寝殿，殿内布局同紫禁城的坤宁宫和沈阳
故宫的清宁宫类似，即东半部为居住空间，西半部为萨满教佛堂。乾
隆时期改建后成为现在的格局，平面布局为十字形。殿面阔七间，单
檐歇山顶，前后各出五间抱厦，殿顶按照藏传佛教传统设有五座小
阁，象征须弥山的五峰。

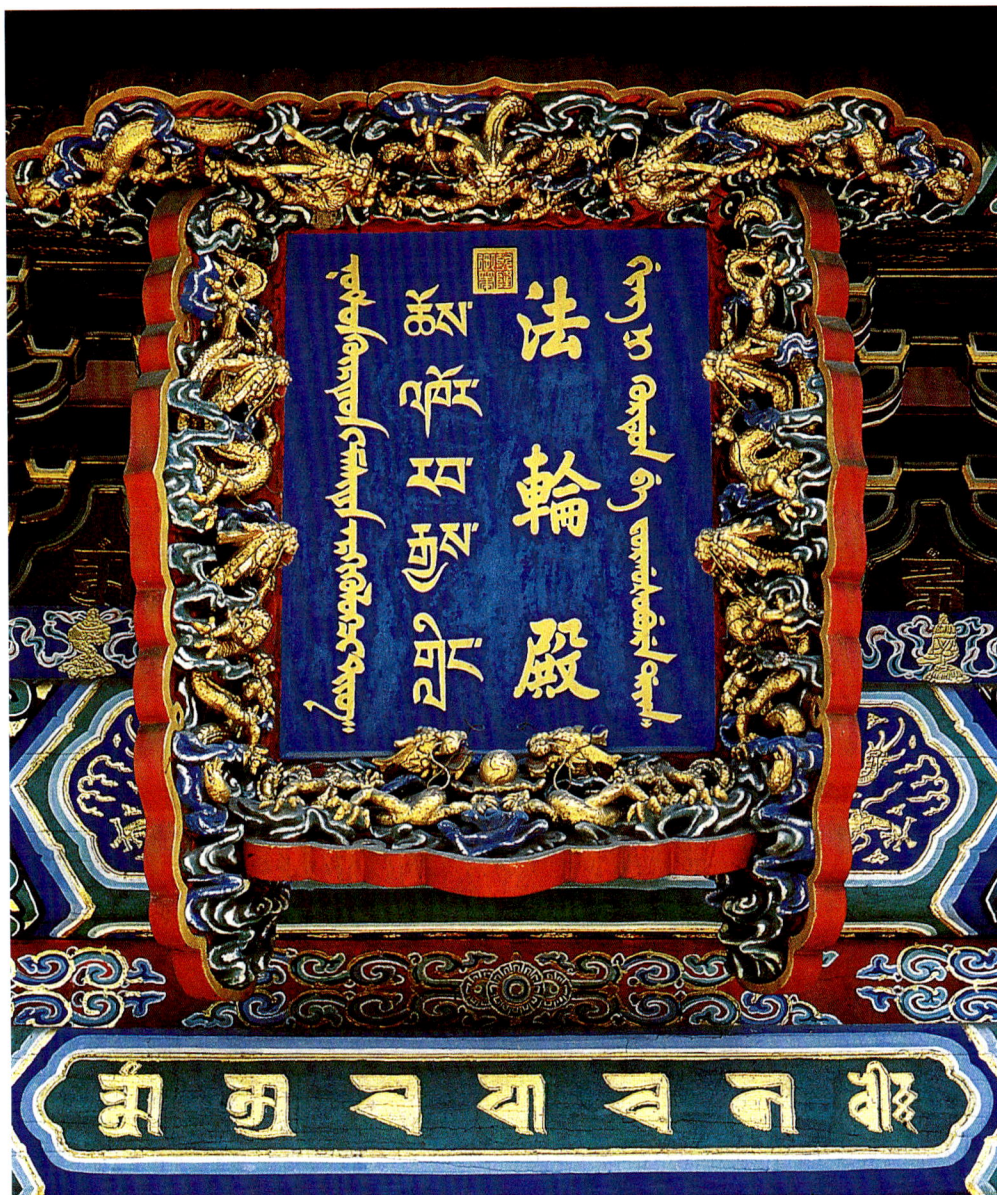

法轮殿匾额
The Board Inscribed with "the Falun Hall"

法轮殿内部供奉
Interior of the Falun Hall

该殿是全庙喇嘛集中念经的地方，殿堂高大。殿正中供奉藏传佛教黄教创始人宗喀巴大师铜像，高6.1米。宗喀巴大师像背后是雍和宫"三绝"之一的五百罗汉山。殿内还有《大藏经》108部、《论藏经》207部及乾隆皇帝亲笔抄写的《药师经》、《大白华盖仪轨经》以及精细的壁画。法轮殿左侧为高阁式的班禅楼（又称药师坛），因1780年六世班禅来京给乾隆皇帝祝寿，在此讲经说法而得名。

万福阁
The Wanfu Pavilion

万福阁又叫"大佛楼"，位于雍亲王府时期的后罩楼位置，是雍和宫内最高的建筑，高30余米，面阔七间，三重檐歇山顶，从阁外表看，是一座三层高楼，但从阁楼里面看，则是一座没有楼板相隔的通体高阁，这是辽金时代的建筑风格。阁两旁是永康阁和延绥阁。三座楼阁之间有飞阁复道凌空连接，宛如仙宫楼阙，是唐朝至辽金时代佛教建筑的典型风格，也是中国现存古建筑中飞阁复道的仅存实例。

万佛阁弥勒大佛
The Huge Buddhist Statue in the
Wanfu Pavilion

大佛像高26米，宽8米，为一整棵
白檀香木精雕而成，为雍和宫"三
绝"之一。

五百罗汉山
The Mountain of Five Hundred Arhats

五百罗汉山是用檀香木细雕精镂而成，高3.4米，宽3.45米，山上林谷、松柏、宝塔、亭阁、洞穴、曲径、小桥流水俱全。五百罗汉是由金、银、铜、铁、锡5种金属铸成，色彩鲜艳，造型生动，是一件质料、造型、雕艺三绝的艺术珍品。可惜历经战乱，山上罗汉仅存449尊了。

旃檀佛
The Sakyamuni Buddhist Statue

旃檀佛是一尊铜胎佛像，供奉于万福阁的东配殿——照佛楼，原为乾隆母亲供佛之处。佛背后有一座围屏式的火焰背光，这火焰背光与佛龛均为楠木雕制，精巧绝伦，被誉为雍和宫的"三绝"之一。

# 潭柘寺
## The Pool and Cudrania Temple

潭柘寺位于北京西郊门头沟区东南部的潭柘山麓，是北京地区有史可考的最古老的寺院，占地面积亦堪称北京寺院之最。2001年被公布为全国重点文物保护单位。

潭柘寺始建于西晋永嘉元年(307年)，名嘉福寺，距今已有近1700年的历史了。唐代时改名龙泉寺，唐代武则天年间扩建和重修。金代皇统年间(1141－1149年)改名大万寿寺，明天顺元年(1457年)复名嘉福寺，清康熙三十一年(1692年)改称岫云寺。其名称更迭频繁，但由于其潭水和柘树非常著名，故俗称潭柘寺。

寺坐北朝南，整体依山势逐层升高，规模宏大，寺内占地2.5公顷，寺外占地11.2公顷。寺分中、东、西三路：中路为主要建筑，最前面为三间四柱三楼牌楼一座，往后依次为山门、天王殿。再往后为位于寺院的中心部位的大雄宝殿，是潭柘寺最重要、建筑等级最高的殿堂。大殿东西两侧对称建有祖师殿和伽蓝殿两座配殿，大殿院内还建有钟鼓楼。大雄宝殿后还有三圣殿(现仅存基址)和毗卢阁。东路建筑是由若干庭院式建筑群组成，主要有大厨房、延清阁、财神殿、方丈院、帝后宫、舍利塔等组建筑，主要是方丈居住、修行和清代皇室进香礼佛时的行宫，环境清幽。西路建筑群是寺院式殿堂的组合，从使用上划分，南侧是法坛式建筑区，主要建筑有举行剃度出家仪式的戒坛、高僧讲经宣法的楞严坛和大悲坛，还有僧人抄写经书的写经室、高等僧人吃饭的南楼等组建筑。北侧是佛殿式建筑区，建筑多依山势而建，主要有观音殿、文殊殿、祖师堂、龙王殿等多组建筑。

寺外有位于潭柘寺左前方用于年老僧人养老的安乐延寿堂；寺院后部东西两侧的东、西观音洞；寺院东部的明王殿建筑群；此外还有潭柘寺从金代起至民国的历代高僧的墓塔群，塔的形式和题记极其丰富，是研究佛教和古建筑的珍贵实物。

潭柘寺附属文物甚多，如镀金鸱带、金代诗碣、清代肉身佛等都是文物珍品。潭柘寺另一奇观就是众多的古树名木，其中国家级保护古树就有186棵之多，最著名的有乾隆皇帝封为"帝王树"和"配王树"的千年白果树及千年娑罗树、千年柏、"百事如意"树。还有明代的二乔紫玉兰、黄玉兰、"金镶玉"和"玉镶金"竹、紫竹、古老的探春花等。这些古老的文物所承载的众多历史的沧桑和许多美妙的传奇故事，更是引人入胜，为这座古老的寺院增添了几分神秘的光环。

潭柘寺全貌
A Panorama of the Pool and Cudrania Temple

Located at the foot of the Pool and Cudrania Mountain in the southeastern part of Mentougou District, the Pool and Cudrania Temple is the oldest and largest temple in Beijing. It was listed as a national key relic under special preservation in 2001.

First built in the 1st year of the Yongjia reign of the Western Jin Dynasty (307 A.D.) and named the Temple of Auspicious Fortune, the temple has existed for more than 1700 years. In the Tang Dynasty, it was expanded and renamed the Dragon Spring Temple. It has been popularly known as the Pool and Cudrania Temple because of the noted Dragon Pool and cudrania trees on the hill behind the temple.

Facing south, the area of the entire temple is more than 11.2 hectares. The spacious buildings stand on the terraces one higher than the other and are divided into the middle, the eastern and the western axes. Along the middle axis, the principal structures, from the front to the rear,

山门
The Front Gate

山门为无梁殿式，面阔三间，单檐歇山顶，灰筒瓦屋面，檐下悬挂康熙皇帝亲书匾额"敕建岫云禅寺"。

are the archway, the front gate, the Hall of Heavenly Kings, the Hall of the Mahavira and the Vairochana Pavilion.

On the eastern axis are several courtyards of building complex. The main buildings are the kitchen, the Yanqing Pavilion, the Mammon Hall, the Abbot Compound, the temporary imperial palace, the dagoba etc. In the courtyard of the imperial palace stands the Floating Cups Pavilion. On the marble floor of the pavilion is a shallow channel shaped like a coiled dragon through which the spring water flows. The pyramidal roof is covered with green glazed tiles, a stele written by Emperor Qianlong hanging under the eave. The western axis is comprised of alter buildings complex and hall buildings complex. The main buildings include the Ordination Altar, the Lengyan Altar, the Altar of Dabei (Great Mercy), the Master Hall, the Hall of the Dragon King and so on.

Another wonder at the temple is numerous ancient and noted trees. On the eastern side of the Mahavira Hall stands an ancient gingko tree known as the Emperor of Trees. The name was given by Qing Emperor Qianlong. There is another symmetrically placed gingko growing on the western side of the hall called the Emperor's Companion Tree. The pines along the middle axis are particularly grand and besides them there are magnolia and Saltrees and a variety of other rare flowers and shrubs.

钟楼
The Bell Tower

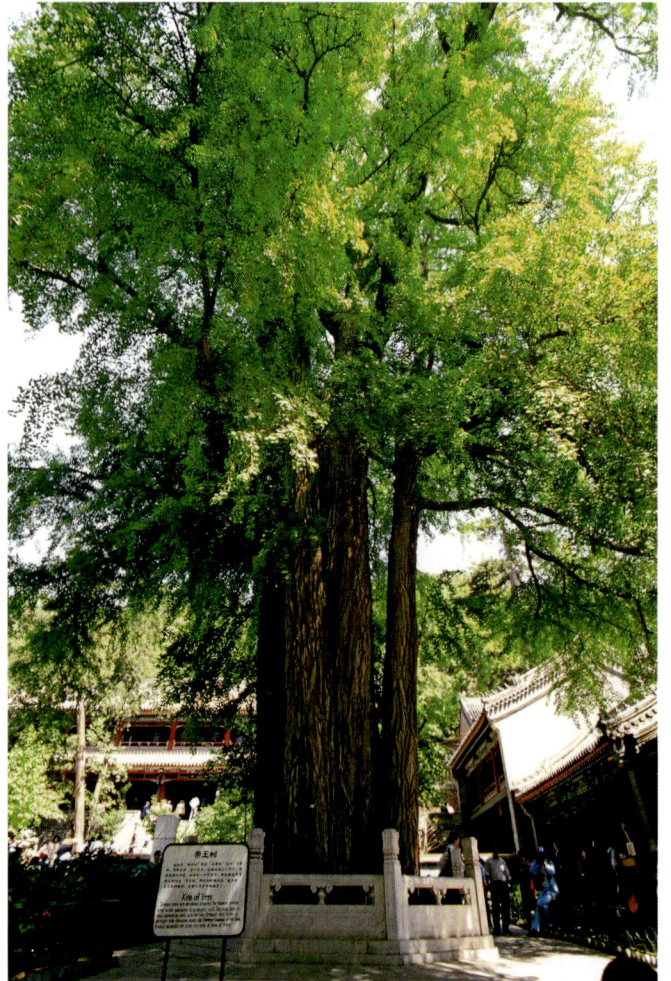

帝王树
The Emperor of Trees

眺望大雄宝殿
A Far Sight of the Hall of the Mahavira

大雄宝殿
The Hall of the Mahavira

大雄宝殿面阔五间，重檐庑殿顶，上层黄琉璃瓦屋面，下层黄琉璃
瓦绿剪边屋面。大殿的正脊高大，正脊两端安放着一对琉璃鸱吻，
高2.9米，鸱吻建造于1692年，是仿照原来元代大吻的样式烧制的，
是北京所有的佛寺中最大的。值得提出的是鸱吻两侧各系有一条金
色锁链，名"镀金剑光吻带"，是康熙皇帝的御赐之物，仅潭柘寺所
用，殿内正中供释迦牟尼像，两侧供奉十八罗汉像。

大雄宝殿内景
Interior of the Hall of the Mahavira

大雄宝殿藻井
The Sunk Panel of the Hall of the Mahavira

流杯亭
The Floating Cups Pavilion

流杯亭是东路帝后宫内的一座比较有特色的建筑，又名"猗玕亭"，是皇室"曲水流觞"的地方，建在原无逸殿的遗址上，单檐四角攒尖顶，绿琉璃瓦屋面，檐下悬乾隆皇帝亲书"猗玕亭"匾。亭内汉白玉石铺地，石面上凿刻有蜿蜒曲折的水槽，宽、深均为10厘米，石槽构成的图案从南向北看为龙头形，从北向南看为虎头形，构思奇妙，所以该亭又叫"龙虎亭"。

流杯亭水槽
The Shallow Channel on the Floor
of the Floating Cups Pavilion

塔林
Pagoda Forest

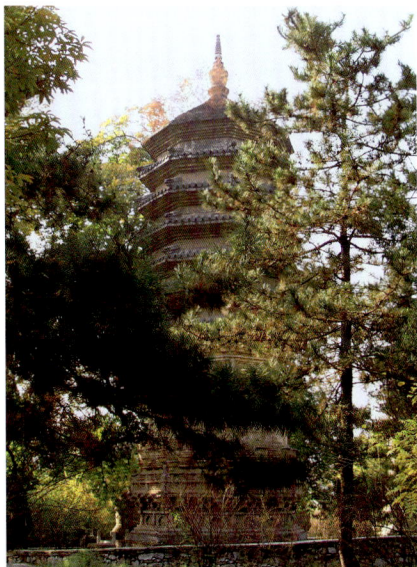

广慧通理塔
The Stupa for Guanghui Tongli Master

舍利塔
The Dagoba

上塔林中覆钵塔
The Pagoda of Upside-Down Alms Bowl in the Pagoda Forest

下塔林中覆钵塔
The Pagoda of Upside-Down Alms Bowl in the Pagoda Forest

下塔林中密檐塔
The Multi-Eaved Pagoda in the Pagoda Forest

# 戒台寺
# Jietai Temple

　　戒台寺位于北京西郊门头沟区马鞍山麓。正名为万寿禅寺。因寺内有一座驰名佛界的大戒坛，故俗称"戒台寺"或"戒坛寺"。1996年被公布为全国重点文物保护单位。

　　戒台寺始建于唐武德五年(622年)，原名慧聚寺。辽清宁年间，高僧法均在寺左创建戒坛传戒。明清时期曾多次修缮和扩建，现存建筑大都是清代所建。

　　寺坐西朝东，依山而建，占地4.3公顷。主要殿堂由两条东西向轴线组成；南侧靠前一组由低逐层升高，山门临第一层高台边缘而建，前有石栏，外门在山门的南侧。山门殿前有石狮一对，清康熙御碑一座，左右墙垣各辟一配门，山门之内，中轴线上天王殿院内钟鼓楼分立左右，楼南各有旗杆座一个，院内四棵古松，明清民国所立石碑数块。左右墙垣上各有便门一个。大雄宝殿院内左右分为伽蓝殿和祖师殿以及厢房，大雄宝殿前有月台，后有石栏、垂带。大雄宝殿后为第二层高台，台上原有千佛阁，已毁，阁后再有两层高台，依次建有三仙殿、九仙殿等。第二层台上于千佛阁之北有戒坛殿。戒坛殿之前为明王殿，实即殿门。门外有经幢三，六角或八角形柱，刻有佛像、经文，其中两座为辽代，一座为元代。它的左右有侧门连墙和戒坛殿左右及后部的庑廊相连，围合成院落，形成寺庙的北侧次要轴线。明王殿前临第二层台的边缘，台下南北并列两座砖塔。南侧五层的八角密檐塔，为辽高僧法均的衣钵塔。北侧为法均的灵塔，塔下的须弥座上还保存着几块辽代的砖雕，但基座以上部分为明正统十三年(1448年)重建。北塔之南有辽大安七年(1091年)王鼎撰文的法均的"遗行之碑"；碑文称法均死于大康元年(1075年)三月，同年五月起灵塔于方丈院之右，可知北塔残基建于1075年，现在塔东侧的空地在辽时为方丈院。此外寺中还有辽金碑、幢数通。在千佛阁遗址与戒坛之间，有牡丹院等庭院式建筑。寺的东南是方丈院，明清著名法师都住在这里。寺的东面是历代和尚的墓地。寺后有极乐峰，峰下有喀斯特溶洞，洞外有佛殿、佛塔遗迹。

　　古刹戒台寺是一座具有悠久历史的寺院，亦是京西大寺之一，也是京郊一处著名的旅游景点。

The Jietai Temple the Ordination Terrace Temple is situated on the foot of Ma'an Mountain in Mentougou District of the city western suburb. Its full name is the Longevity Temple, commonly known as Jietaisi. The temple takes its name from its ordination altar. It was listed as a national key relic under special preservation in 1996.

The temple was first built in the 5th year during the Wude reign of the Tang Dynasty (622 A.D.) and named

Huijusi (the Wisdom Accumulation Temple). During the Qingning reign of the Liao Dynasty, a monk named Fajun had an altar built here for the ordination of novices into the Buddhist priesthood. Then, the temple was renovated and enlarged several times during the Ming and Qing Dynasties. Most of the buildings date back to the Qing Dynasty.

The temple, which faces east, occupies 4.3 hectares of land. The main buildings are laid out orderly by two eastern and western axes. Along the southern axis are the front gate, the Hall of Heavenly Kings, the Mahavira Hall, and the Pavilion of 1,000 Buddhas (the existent historic site). On the northern axis lie the Mingwang Hall and the Altar Hall. The temple also contains a Liao Dynasty stupa and a Jin Dynasty stela.

戒台寺全貌
A Panorama of the Ordination Terrace Temple

卧龙松
The Reclining-Dragon Pine

观音殿
The Hall of the Kwan-Yin

大雄宝殿
The Hall of the Mahavira

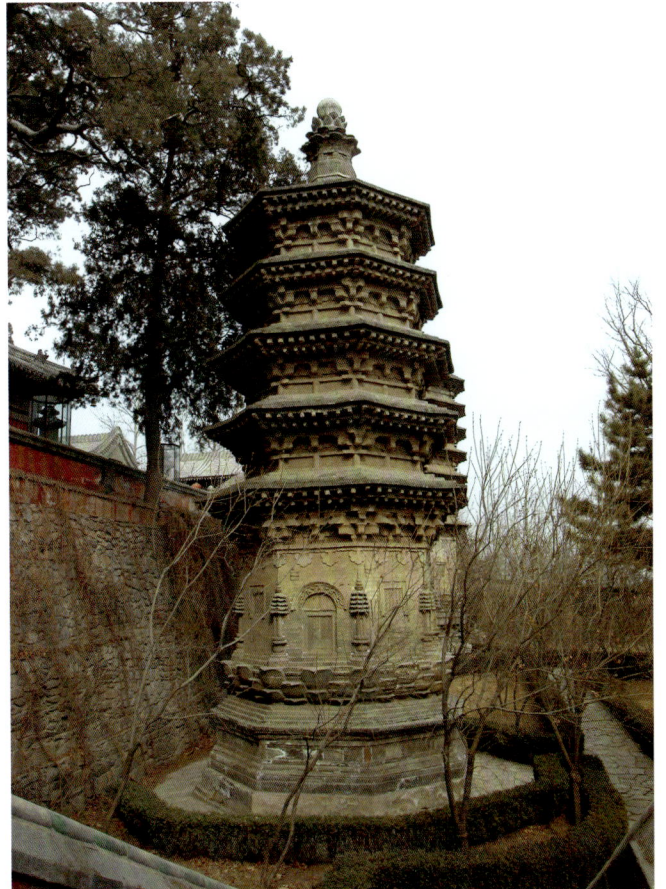

辽塔
The Stupa of the Liao Dynasty

法均的衣钵塔，塔身八面都有浮雕，四正面用砖雕砌格子门，四斜面为琐纹窗，八个转角各有一座附壁的五层檐密檐小塔。各层角梁都是砌在砖檐之内的木梁上。此塔主要部分还是辽代的原式，仅塔下须弥座和塔刹已经改动过。

戒台殿
The Ordination Terrace Hall

戒台殿为二层攒尖屋顶的建筑，下层方五间，上层方三间，殿内有高三层的白石戒坛，为明代遗物。戒坛是一个高丈余的三层石刻台座，雕刻精美，是我国现存戒坛中最大的一座。坛周围原有数百个戒神；殿堂内有明代沉香木雕花椅，系传戒时三师七证的座位。殿外有金主完颜亮时所立的传戒大师碑。

方丈院
The Abbot Compound

牡丹院
The Peony Courtyard

戒坛
The Ordination Altar

乾隆曾带后妃在此避暑，后为恭亲王奕䜣的常住之所。

# 云居寺塔及石经
# Yunju Temple

云居寺位于北京市房山区尚乐水头村，由寺院、塔群和石经山藏经洞三部分组成。1961年房山云居寺塔及石经被公布为全国重点文物保护单位。

云居寺始建于隋代末年，为隋代高僧静琬法师创建，寺院宏大，古塔屹立，而且其石经、木经、纸经更是被誉为三绝。云居寺是研究北京历史的宝贵实物资料。鉴于前代的"灭佛"事件，静琬法师开始在石经山上把经书刻在石板上以求长期保存，由静琬法开创的刻经事业历经隋、唐、辽、金、元、明六个朝代，绵延1039年，镌刻佛经1122部、3572卷、14278块，成为云居寺的一大佛教文化宝库。伴随着刻经，云居寺成为佛教胜地，一直兴盛不衰。抗日战争期间，寺院殿堂被日军全部炸毁。建国后，政府又重新修复了殿堂，恢复了昔日的风貌。

云居寺坐东朝西，依山而建，渐次升高，中轴线上有六进殿宇，依次为天王殿、毗卢殿、释迦殿、药师殿和弥勒殿。云居寺的最高处是大悲殿，它与说法堂、藏经阁构成全寺规模最大的殿宇。寺北有一座辽代砖塔，名罗汉塔，俗称北塔，塔四周还有四座唐代密檐小塔；寺南亦曾有一座砖塔，叫压经塔，俗称南塔，可惜已毁于抗日战争期间的炮火中。云居寺除北塔和南塔外，石经山上还有不少大小佛塔。尤其一座建于唐太极元年(712年)的唐塔，堪称北京地区塔的元老。这些塔是研究古代佛学及建筑学的极好实物。

云居寺最为著名的是石经，它是中国唯一的石刻《大藏经》。在石经山半山腰，开凿有9个藏经洞，分上、下两层，其中8个洞为封闭式，每洞装满经版后，用石堵门，以铁水浇铸，牢固异常。只有规模最大的雷音洞为开放式，静琬最初刻经146块，都嵌在这个洞的四壁，洞内有4根八面的主柱，柱上雕有佛像1056尊，故称千佛柱。九个洞内共藏经石4195块。山下南塔，塔下藏经石10083块。堪称世界杰作。石经不仅是我国古代佛教典籍的铭刻，也是世界佛经铭刻之最。

Located at Dashiwo Town in Fangshan District, the Yunju Temple consists of three parts: monastery courtyards, pagoda forest and sutra storage caves in the sutra hill. It was listed as a national key relic under special preservation in 1961.

The temple was first built during the last years of the Sui Dynasty by the eminent monk Jingwan, who was the first monk engraving sutras on flagstones. The carving work started in the Sui Dynasty, and lasted 1039 years through six dynasties the Sui, Tang, Liao, Jin, Yuan and Ming. The 14,278 stone slabs contain 1,122 Buddhist scriptures in 3,572 volumes. So, the temple was called a treasure house of ancient Buddhist culture. During the Period of Anti-Japanese Aggression, all the halls of the temple were blown down by Japanese army. After the People's Republic of China was founded, the halls were restored and renovated with funds from the government.

Facing west, the spacious buildings stand on the terraces one higher than the other. The temple comprises six courtyards one behind another in the central axis. The principle buildings, from the front to the rear, are the Hall

山门
The Front Gate

of Heavenly Kings, the Hall of the Sakyamuni, the Hall of the Bhaisajyaguru, and the Hall of the Maitreya. To the north of the temple stands a brick pagoda, the Arhat Pagoda, built in the Liao Dynasty, also known as the Northern Tower, which is surrounded with 4 multi-eaved pagodas built in the Tang Dynasty.

The Yunju Temple is famed for its stone slabs inscribed with Buddhist scriptures. Located on the mountainside of the sutra hill, nine sutra storage caves are divided into the upper and the lower parts, where hide 4,195 pieces of stone slabs. Among them, only the Leiyin Cave is open to the public. 146 stone slabs inscribed with Buddhist scriptures carved by Jingwan are kept in the cave and all of them are mounted on the walls of the cave. In the cave, there are four stone pillars carved with 1,056 Buddhist images, also known as thousand-Buddha pillars.

牌楼
The Archway

藏经洞外景
Exterior of Sutra Storage Caves

雷音洞石经
Stone Slabs Inscribed with Buddhist Scriptures in the Leiyin Cave

雷音洞
The Leiyin Cave

唐塔
The Stupa of the Tang Dynasty

唐塔内壁雕刻佛像
Buddhist Carvings on the Inner
Wall of the Tang Dynasty Stupa

北塔佛偈语刻砖
Brick Carvings of the Northern Pagoda

# 碧云寺
# The Temple of Azure Clouds

　　碧云寺位于北京香山东麓，是北京西山著名的寺院之一，寺内环境清幽，建筑高低错落，金刚宝座塔和五百罗汉像更堪称一绝。2001年被公布为全国重点文物保护单位。

　　寺始建于元朝至顺二年(1331年)，元丞相耶律楚材之后裔耶律阿吉舍宅为寺，初名碧云庵，后改碧云寺。明清两代均有扩建。明正德年间太监于径在此大兴土木，为第一次扩建，并在寺后为自己修建坟墓。嘉靖初年于径获罪，不能在此处安葬。天启年间太监魏忠贤又扩建庙宇，再次建坟，准备死后葬此。崇祯初年，魏忠贤自缢后被戮尸，也不能再葬于此。魏忠贤的党羽葛九思，随清军入京，将魏之衣冠葬在墓中，成为魏的衣冠冢。直到康熙四十年(1701年)，江南道监察御史奉命巡视西山时，得知是魏坟，遂于五月十二日上奏，二十二日诏平其坟。乾隆十三年(1748年)，对寺宇重加修葺，并按西僧所贡奉的图样，建起了金刚宝座塔，同时新建了行宫和罗汉堂，对其他殿宇无大变动。所以寺内殿宇基本是明代结构。1925年孙中山先生在京逝世后，曾在该寺的后殿停过灵柩，因而此殿后改为中山堂。金刚宝座塔下，成为孙中山衣冠冢。

　　碧云寺坐西朝东，分为中路主要建筑、水泉院建筑和五百罗汉堂建筑三大部分。

　　中路建筑为主要殿堂所在，最前方为一座石桥，东桥头有明代雕制的汉白玉石狮一对。山门位于桥后，为入寺庙的第一道门，以后依次为山门殿三间、弥勒殿、大雄宝殿、菩萨殿、孙中山纪念堂和金刚宝座塔。水泉院位于碧云寺北跨院，是一座融园林景观和居住殿堂相结合的皇家行宫建筑，整个院落坐西朝东，前三进为含青斋建筑群、后两进为水泉院。

Situated in the eastern foot of Xiangshan Mountain of Beijing, the Temple of Azure Clouds is known as one of the most famous temples in the Western Hills. The temple buildings are scattered on the slopes uprising gradually and from the temple's main gate at the foot of the hill to the highest point, rise almost 100 meters. It was listed as a national key relic under special preservation in 2001.

The temple was first built in the 2nd year during the Zhishun reign of the Yuan Dynasty (1331 A.D.) and named the Nunnery of Azure Clouds. In the Ming Dynasty, two powerful eunuchs, Yu Jing and Wei Zhongxian had it expanded at various periods, trying to make it their burial ground. In the 13th year of the reign of Emperor Qinglong (1748 A.D.), large-scale construction work was done. The Diamond Throne Pagoda designed according to the drawing provided by western monk was built at the rear of the temple, the Hall of Arhats

and temporary imperial palace were also built in this period. When Mr. Sun Yat Sen died in 1925, his coffin was briefly on display here, but a permanent memorial hall remains at the temple.

The temple, which faces east, comprises three building complexes. There is a stone bridge spanning right in front of the temple with two huge statues of lions. The principal structures, from the front to the rear, are the front gate, the Hall of Heavenly Kings, the Hall of the Mahavira, the Hall of Bodhisattvas and the Sun Yat Sen Memorial Hall. At the rear of the temple is the Diamond Throne Pagoda.

碧云寺全貌
A Panorama of the Temple of Azure Clouds

山门
The Front Gate

山门建筑在高台上，为无梁殿形式，灰筒瓦屋面，山面各连一座碉楼
式建筑，两层，单檐歇山顶，灰筒瓦屋面，虎皮石墙面。

金刚宝座前石牌坊
The Stone Archway in Front of the Diamond Throne Pagoda

碧云寺石牌坊枋心忠孝廉节人物像
Figures Carved on the Inner Screen Walls of the Stone Archway in the
Temple of Azure Clouds

碧云寺石牌坊枋心披发母狮
Maned Female Lion Carved on the Back of the Right Inner
Screen Wall of the Archway in the Temple of Azure Clouds

大雄宝殿
The Hall of the Mahavira

大雄宝殿位于弥勒殿后，面阔三间，单檐庑殿顶、灰筒瓦屋面、回廊，檐下旋子彩画，后檐明间出抱厦一间，旋子彩画。殿内主佛供奉释迦牟尼，殿内天花藻井十分精美。殿前月台上南北经幢各一座、南北配殿各三间。

金刚宝座塔
The Diamond Throne Pagoda

金刚宝座塔在中路的最后面，塔前建木牌楼门、石牌坊、砖石牌坊作为前导，更衬托出金刚宝座塔的神秘与巍峨。

塔通高34.7厘米，下部砌两层虎皮石基座，其上建塔。塔全部用华丽的汉白玉砌成，分上下两部分。下为金刚宝座，正面正中开券洞，孙中山的衣冠即封葬于此。循石阶登上塔座，出口处置一方亭，左右各有一藏式覆钵塔，后为五座十三层密檐方塔，中央为主塔，四隅各建子塔。整个金刚宝座塔满布雕刻精美的浮雕，有大小佛像、天王、力士、龙凤狮象和云纹梵花等，依西藏传统形式雕刻，是乾隆年间的石雕精品。

金刚宝座上覆钵式小塔
The Sub-Pagoda of Upside-Down Alms Bowl of the Diamond Throne
Pagoda

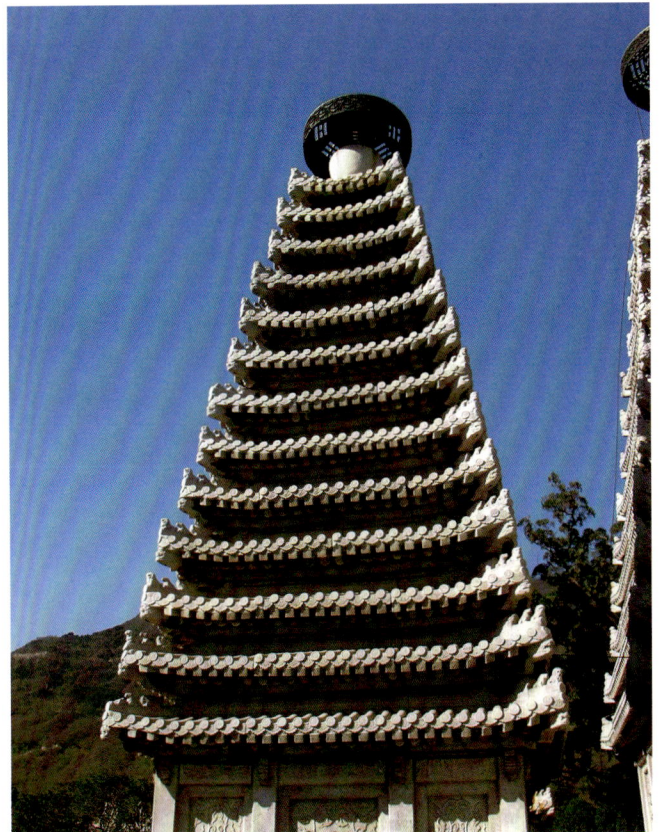

金刚宝座上密檐式小塔
The Multi-Eaved Sub-Pagoda of the Diamond Throne Pagoda

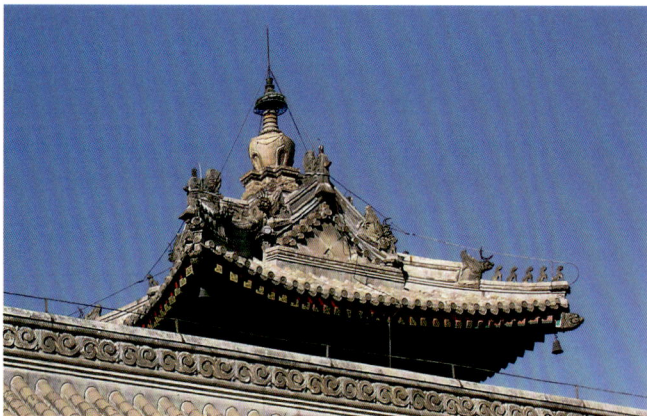

罗汉堂屋面
The Roof the Hall of Arhats

罗汉堂
The Hall of Arhats

罗汉堂建于乾隆十三年(1748年)，仿杭州净慈寺罗汉堂而建，平面呈"田"字形，每面九间，灰筒瓦，庑殿顶，中间夹有4个天井，十字交叉处建重檐方亭，正吻处为一覆钵式塔，四面入口处建抱厦，前檐明间檐下金字卧匾书"罗汉堂"，正门内供奉四大天王，室内有罗汉像508尊。罗汉形态各异，生动传神，表现了高超的雕塑技艺，是碧云寺的一绝。传说康熙、乾隆也进入罗汉之列。

殿中心为三世佛像，四面通道上各立一尊佛：东为护法金刚韦驮，北面为疯僧，西面是地藏菩萨，南为接引佛。北面房梁上有济公活佛。

罗汉堂内五百罗汉
Five Hundred Arhats in the Hall of Arhats

# 万寿寺
# Wanshou (Long Life) Temple

万寿寺位于北京市海淀区西三环中路紫竹桥东北,是一座融寺院建筑和明清两代皇家行宫为一体的寺庙。2006年被公布为全国重点文物保护单位。

万寿寺始建于唐朝,称聚瑟寺。明万历五年(1577年)重修,改名万寿寺。清康熙、乾隆年间三次重修,并增建了西路行宫建筑。光绪初年毁于火,光绪二十年(1894年),为给慈禧祝寿再次重修万寿寺,在西跨院增修千佛阁和梳妆楼,成最后格局。光绪二十六年(1900年)再次修缮寺院。

寺分三路,中路为主体,西为行宫,东为方丈院,共占地31800平方米。中路有七重殿,入山门的前四进殿由南而北依次为:天王殿、大雄宝殿、万寿阁、大禅堂;第五进院落很有特色,建筑置于假山之上,中为观音殿、东为文殊殿、西为普贤殿,寓意中国佛教普陀、清凉(五台山)、峨眉三仙山;院中还有一座乾隆御碑亭。第六重殿为无量寿佛殿,中路的最后一座建筑是万寿楼。西路的主要建筑有大门、寿茶房、寿膳房、前正殿、正殿、梳妆楼、大悲堂等建筑,建筑既有宫廷风格又有园林气息,慈禧游颐和园经常在这里歇息、喝茶。

万寿寺全貌
A Panorama of the Wanshou Temple

Located at the northeast of Zizhu Bridge in the western 3rd Ring Road, the Wanshou (Long Life) Temple is a well-known ancient temple in the western part of the city, also a combination of temple and temporary imperial palace in the Ming and Qing Dynasties. It was listed as a national key relic under special preservation in 2006.

The temple was first built during the Tang Dynasty and named Jusesi. It was reconstructed and renamed Wanshousi in the 5th year of the reign of Emperor Wanli of the Ming Dynasty (1577 A.D.). During the reigns of Emperor Kangxi and Qianlong, it was rebuilt three times and the temporary imperial palace was added in the western section. The temple was renovated in the 20th year of the reign of Emperor Guangxu (1894 A.D.) for birthday celebration of Empress Dowager Cixi.

The temple covering an area of 31,800 square meters is divided into the middle, the eastern and the western axes. The principal structures, from the front to the rear, on the middle axis, are the Hall of Heavenly Kings, the Mahavira Hall, the Wanshou Hall, the Kwan-yin Hall, the Hall of the Amitayus, the Wanshou Building and so on.

大雄宝殿
The Hall of the Mahavira

大雄宝殿为单檐庑殿顶琉璃瓦屋面，殿内供奉"三世佛"及毗卢佛，东西配殿各三间，为歇山顶筒瓦屋面，东曰"祝延万寿"，西曰"安心镜"。

三大士殿
The Sandashi Hall

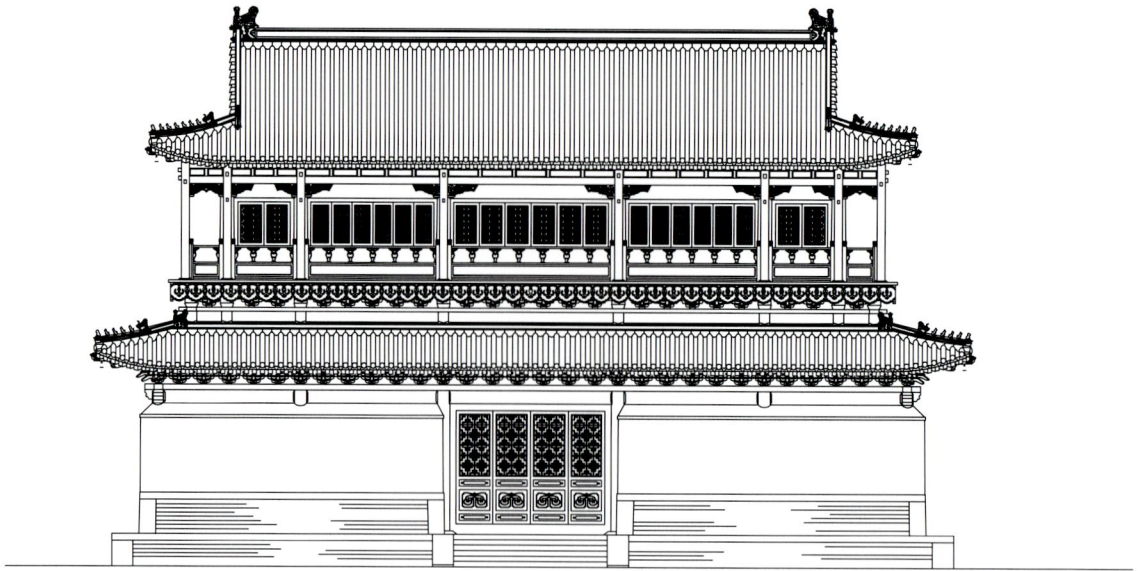

万寿寺万寿阁北立面图
The Elevation of the Wanshou Pavilion of the Wanshou Temple

万寿阁
The Wanshou Pavilion

万寿阁又名宁安阁，面阔五间，重檐歇山顶筒瓦屋面，东西配殿各三
间，东为韦驮殿，西为达摩殿，均为歇山顶筒瓦屋面。

中西合璧式院门
The Courtyard Gate in a Combination of Chinese and Foreign Styles

中西合璧式院门位于无量佛殿两侧，为巴洛克式，建于清乾隆二十六年(1761年)，在皇家寺院中风格独特。

中西合璧式院墙(局部)
Section of the Courtyard Gate in a Combination of Chinese and Foreign Styles

乾隆御碑亭
Emperor Qianlong's Imperial Tablet Pavilion

乾隆御碑亭位于第五进院落，为重檐八角攒尖顶，黄琉璃瓦屋面，亭中为乾隆题"重修万寿寺"碑，碑文用汉、满、蒙、藏四种文字。万寿寺的无量寿佛殿后还有光绪御碑亭一座，形制与此碑亭相同，亭内为光绪二十年(1894年)翁同龢书的碑文。

西路垂花门
The Drooping Flowers Gate on the Western Axis

慈禧梳妆楼彩绘
Coloured Paintings of the Storied Building for
Empress Dowager Cixi to Dress and Make up

慈禧梳妆楼
The Storied Building for Empress Dowager Cixi to Dress
and Make up

**67**

# 十方普觉寺(卧佛寺)
# The Temple of Universal Awakening

卧佛寺在西山余脉聚宝山(又名寿安山)的南麓，背倚山冈。2001年被公布为全国重点文物保护单位。

寺首创于唐贞观年间(627—649年)，原名兜率寺，以后历代有废有兴，寺名也屡经更改，清雍正十二年(1734年)改称十方普觉寺，沿用至今。因寺内有元至顺二年(1331年)所铸一座释迦牟尼像，俗称卧佛寺。现存建筑均为清代所建。

寺坐北朝南，建筑规整，由三组院落组成。寺前有一座四柱三楼木牌坊，额题"智光重朗"，北额"妙觉恒玄"，其后百余米长的坡道上巨大的黄琉璃牌坊为外门，额书"同参密藏"，背额"具足精严"，均为清乾隆帝御书。坊后有半圆形水池及桥，桥北为山门殿，左右有钟鼓楼。山门左右有侧门，门外侧连廊庑，至角转向北，一直延伸至最后一进建筑藏经楼的两侧，围成纵长的殿庭，这种廊院是元代以前的布局风格。在殿庭的中轴线上有山门、天王殿、三世佛殿、卧佛殿、藏经楼。东、西两路在中路两侧，都由前后几进院落组成，为清代的行宫和附属殿堂，均为复建。卧佛寺中路殿亭前、中、后三重殿宇，东西路各建若干院落，与中路间隔有南北巷道，多少还反映出一些唐宋巨刹在东西廊外分列各院的廊院制度的遗意，对了解古代卧佛寺的发展演变是有参考价值的实物资料。

The Temple of Universal Awakening is located at the southern foot of Jubao (also called Shou'an) Mountain in the Western Hills. The rear of the temple is set against the mountain cliffs. It was listed as a national key relic under special preservation in 2001.

The temple was first built during the Zhenguan reign of the Tang Dynasty (627-649 A.D.) and originally named Doushuaisi. It was renamed the Temple of Universal Awakening in the 12th year of the reign of Emperor Yongzheng of the Qing Dynasty (1734 A.D.). It is generally referred to as the Temple of the Reclining Buddha for the reason that there is a Buddha made in the 2nd year during the Zhishun reign of the Yuan Dynasty (1331 A.D.).

The temple, which faces south, comprises three courtyards. There is a yellow glazed archway standing right in front of the temple. In the centre of the first courtyard is a little pond with a stone bridge spanning on it. The Bell Tower and the Drum Tower stand respectively on each side of the courtyard. On the middle axis, from the south to the north, are the front gate, the Hall of Heavenly Kings, the Hall of Buddhas in the Three Periods, the Hall of the Reclining Buddha and the two-storied Sutra Hall.

琉璃牌坊石额
The Stone Board of the Glazed Archway

牌坊琉璃装饰
The Glazed Decoration on the Glazed Archway

琉璃牌坊
The Glazed Archway

三世佛殿
The Hall of Buddhas in the Three Periods

三世佛殿面阔五间，进深三间，绿琉璃瓦黄剪边单檐歇山顶，殿额为雍正御书"双林邃境"，殿内供奉三世佛、倒座观音与十八罗汉像，殿外东有配殿，供有"悉达多太子"和"波斯匿王"像，西配殿供奉"达摩祖师"和"地藏菩萨"像。殿左有雍正十二年(1734年)《御制十方普觉寺碑》，殿右为乾隆五十二年(1787年)《重修十方普觉寺落成瞻礼诗》碑。

东路后罩楼
The Shielding Building on the Eastern Axis

卧佛
The Reclining Buddha

卧佛殿面阔三间，绿琉璃瓦黄剪边单檐庑殿顶，门额书"性月恒明"，侧楹联为慈禧书写，殿内檐悬乾隆御笔"得大自在"。殿中铜卧佛长5.3米、高1.6米、重54吨，据传铸造于元代至治元年（1321年）。

# 妙应寺白塔
# The White Stupa in the Miaoying Temple

妙应寺位于北京市西城区阜成门内大街路北，因寺内有通体涂以白垩的佛塔，故俗称白塔寺。1961年被公布为全国重点文物保护单位。

辽代曾在此建舍利塔，后毁于战火。元至元八年（1271年），元世祖忽必烈令尼泊尔工艺家阿尼哥主持在辽塔遗址上修建了这座大型喇嘛塔，于至元十六年竣工，并迎释迦牟尼舍利藏于塔中。同年，忽必烈又命以白塔为中心，修建了规模宏大的寺院，赐名"大圣寿万安寺"。元末，寺院全部殿堂被雷火焚毁，仅白塔幸免。明天顺元年（1457年），英宗下诏复建寺院，赐名"妙应寺"。清康熙、乾隆、嘉庆年间都数次对寺和佛塔进行了修葺。

寺由寺院和塔院两部分组成。寺院部分共分为三组院落，有天王殿院、三世佛殿院、七佛宝殿院。每座院落相互独立又相互连通，建筑布局严谨完整。主要建筑全部建于南北的中轴线上，主殿的两侧均建有附属建筑，左右对称。塔院建于全寺的最后面，自成一封闭院落。塔院地基高出前面寺院建筑2米左右，院内四角各建角亭一座。白塔位于塔院的中央，是全寺中最高大的建筑，通高51米，也是北京保留规模最大的覆钵式佛塔。白塔由塔座、塔身、相轮、华盖和宝顶组成。塔院主殿为"具六神通"殿，殿内供奉三世佛，殿后北墙建有佛灯龛3座，龛内砌成台阶形式，其上安放供佛油灯。

The Miaoying Temple is located to the north of the Fuchengmennei Street. The stupa is like a huge upside-down alms bowl, which is the biggest of this kind in Beijing. However, the temple is known by its popular name the White Stupa Temple. The temple and its stupa were listed as a national key relic under special preservation in 1961.

The Lama stupa was built by the order of Kublai Khan, the founder of the Yuan Dynasty and was designed by a Nepalese architect and technologist Arniger, in the 8th year of the Zhiyuan reign of the Yuan Dynasty (1271 A.D.). It was completed 8 years later. The great temple surrounded with the stupa was built by the order of Kublai Khan and named Dashengshouwan'ansi in the 16th year of the Zhiyuan reign of the Yuan Dynasty. Emperor Yingzong issued an imperial edict to rebuild the temple and named it as Miaoyingsi in the 1st year of the reign of Emperor Tianshun of the Ming Dynasty (1457 A.D.). The temple and its stupa were renovated many times during the Qing Dynasty.

The temple consists of two parts: monastery courtyards and the stupa courtyard. The monastery comprises three courtyards one behind another. The principal structures are the Hall of Heavenly Kings, the Hall of Buddhas in the Three Periods and the Hall of Seven Buddhas, all standing in the central axis and flanked by additional buildings. The stupa courtyard is situated at the northern end of the temple. There are four corner towers on the four corners. In the central section of the courtyard, a 51-meter-high white stupa is the highest building in the temple. It contains three sections: the base, the body and the spire.

山门
The Front Gate

天王殿
The Hall of Heavenly Kings

大殿
The Main Hall

白塔
The White Stupa

白塔塔座高9米，共分三层，最下层为护墙，平面呈方形，台前有一通道，前设台阶，可直登塔基。中层和上层是折角须弥座，平面呈"亚"字形，四角均向内递收二折。上层须弥座上匝放有铁灯龛108座。须弥座上为覆莲座，莲座外有5道环带形金刚圈，用以承托塔身。塔身是一巨型覆钵体，直径为18.4米，外形雄浑光洁，塔体环绕7条铁箍，使塔身形成一个坚固的整体，这是保证白塔永固的重要技术保障。塔身之上又是一层折角式须弥座，是连接塔身和相轮的过渡载体，外形十分自然得体。相轮又名十三天，是象征着佛界的十三层天。白塔的相轮共13层，底大上小，呈圆锥状，层层拔起承托华盖。外观洁白高耸，十分美丽壮观。华盖又称天盘，位于白塔的顶部，直径9.7米，其构造是木制圆盘，外包铜板，盘上部做成房屋常见的筒

瓦形式，便于排水。铜盘周围悬挂36片铜质透雕佛和梵文字体的"华曼"流苏，每片流苏端部悬挂风铃一枚，共36枚，微风吹动，铃声悦耳，每枚风铃上都刻有善男信女的名字和捐资的数量。华盖的中央部位是白塔的塔刹，是白塔最高的建筑构件，是一座高5米、重达4吨的鎏金铜质小塔，其形制与白塔形制基本相同，也有覆钵体和相轮，为保证塔刹的稳固，以8条粗壮的铁链将宝顶固定在铜盘之上。1978年对白塔进行了维修加固工程中，发现了清代乾隆十八年(1753年)存留在高塔顶部的一批珍贵佛教文物，其中有724函的龙藏新版《大藏经》、乾隆皇帝手书经咒、铜三世佛像、赤金舍利长寿佛、雕观世音像、补花袈裟、五佛冠及数十粒舍利子等，内容丰富至极，为镇寺的国宝。

宝顶
The Spire of the White Stupa

# 法源寺
# Fayuan Temple

法源寺位于北京市宣武区广安门内路南的法源寺前街7号。1979年被公布为北京市文物保护单位。2001年被公布为全国重点文物保护单位。

法源寺始建于武则天执政的万岁通天元年(696年)，初名"悯忠寺"，辽金以来一直是著名的古刹。元明之际寺被毁，明正统二年(1437年)由太监出资重建，改名崇福寺。现在寺的规模就是那时形成的，面积比唐、辽时缩小了一半以上。清雍正十二年(1734年)大规模修缮后，改名律宗庙。乾隆四十三年(1778年)，再次重修改名为法源寺，乾隆皇帝御笔亲题"法海真源"。寺内现存建筑基本上都是清代改建的。

寺庙坐北朝南，自山门向北有一个长50米左右的庭院，由六进院落组成。中轴线上的主要建筑有山门、钟鼓楼，北面正中是天王殿。天王殿左右有侧门连廊庑，廊庑又向北折，直抵最后一进建筑藏经阁，形成一个东西50米，南北180米的封闭殿庭。殿庭内正中从南往北依次为大雄宝殿、戒坛、无量殿、大悲坛和藏经楼。中轴线的东西两侧还建有数层院落，为方丈院和僧众的居住场所及寺院用房。每进院落相对封闭而又互相连通，布局严谨整齐。

唐时的悯忠寺在前部曾有东西两座塔，东塔由安禄山造，西塔为史思明造。塔久已毁去，只有史思明造西塔时的铭刻现在还保存在寺中，证明该寺具有悠久的历史。另外在大雄宝殿门间的前金柱下还有两个覆莲柱础，雕工精丽，饱满圆润，至迟是辽代的，也有可能即是唐代所遗。

1979年，中国佛教图书文物馆在法源寺成立，成为收藏、研究和展览佛教史物、图书及佛教教义和佛教史的重要机构。

Faynan Temple (the Temple of the Origin of the Dharma) is located at No.7 Fayuansiqian Street in Xuanwu District. It was listed as a Beijing's relic under preservation in 1979 and was upgraded as a national key relic under special preservation in 2001.

It was first built in 696 and named Minzhongsi (the Temple in Memory of the Loyal) in Tang Dynasty, which was the noted temple in the Liao and Jin Dynasties. During the last years of the Yuan Dynasty and the early years of the Ming Dynasty, the temple was destroyed. In the 2nd year of the reign of Emperor Zhengtong of the Ming Dynasty (1437 A.D.), eunuchs allocated funds to rebuild the temple and renamed Chongfusi (the Temple of Exalted Happiness). The scale of the temple was formed at that time. However, the area was reduced by half of that in the

山门
The Front Gate

天王殿
The Hall of Heavenly Kings

Tang and Liao Dynasties. The temple was reconstructed and renamed Fayuansi in the 43rd year of the reign of Emperor Qianlong of the Qing Dynasty (1778 A.D.). Now most structures in the temple were rebuilt in the Qing Dynasty.

The temple, which faces south, comprises six courtyards one behind another. On the middle axis, from the front to the rear, are the front gate, Drum and Bell Towers, the Hall of Heavenly Kings, the Hall of the Mahavira, the Altar of the Abstinence, the Hall of the Amitayus, the Hall of Dabei (Great Mercy) and the two-storied Sutra Hall. In 1979, the China Buddhism Library was established at the temple, which became an important institution of collecting, researching and exhibiting Buddhism cultural relics, books, doxies and history.

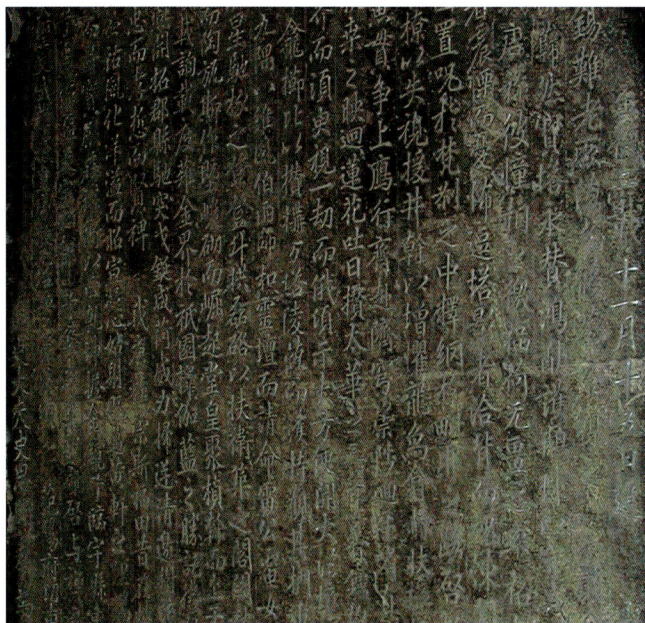

唐代史思明碑刻(局部)
Section of Shi Siming's Stele of the Tang Dynasty

悯忠殿
The Hall in Memory of the Loyal

毗卢殿内
Interior of the Vairochana Hall

大雄宝殿
The Hall of the Mahavira

唐代柱础
Pillar Base of the Tang Dynasty

# 大觉寺
# Dajue Temple

　　大觉寺位于北京市海淀区北安河乡的旸台山麓。1979年被公布为北京市文物保护单位。2006年被公布为全国重点文物保护单位。

　　大觉寺创建于辽咸雍四年（1068年），初名为清水院，金代时为金章宗西山八院之一，后改称灵泉寺。明宣德三年（1428年），由国家出资扩建重修，改名大觉寺。正统、成化年间，又进行了两次较大规模的重修。明末寺宇建筑倾圮。清康熙五十九年（1720年），对寺院进行了一次大规模的修建。清乾隆十二年（1747年），又重修了主要殿宇，并在大悲坛后为该寺住持迦陵和尚修建灵塔（舍利塔）。20世纪30年代，山门被雷击毁。

　　寺坐西朝东，占地面积9500平方米。建筑布局分三路，均依山势修建，保留了辽代寺庙"朝日"的习惯。

　　从山门入寺，经石桥，中轴线上布置着天王殿、大雄宝殿、无量寿佛殿、大悲坛。三座主殿及配殿以大雄宝殿为中心形成一个封闭的空间。经大悲坛，沿山势而上，空间豁然开朗，为一园林式庭院，园中有清代造型玲珑的舍利塔，其旁环抱松柏。寺中轴线的尽头是两层硬山木结构的龙王殿。殿前有八角形汉白玉护栏的水池，池中晶莹的泉水沿阶而下。

　　相传寺内龙潭水系、池中水兽为辽代遗物，而天王殿、无量寿佛殿、龙王殿为明代遗物，憩云轩等则为清代建造。

山门
The Front Gate

Located at the foot of Yangtai Mountain in Haidian District of Beijing, the Dajue Temple was listed as a Beijing's relic under preservation in 1979 and was upgraded as a national key relic under special preservation in 2006.

Construction of the temple began in the 4th of the Xianyong reign of the Jiao Dynasty(1068 A.D.). It was originally one of eight temples in the Western Hills region during the reign of Emperor Zhangzong of the Jin Dynasty, called Qingshui Court, Later, it was renamed Lingquan Temple. In the 3rd year of the reign of Emperor Xuande of the Ming Dynasty (1428 A.D.), the state allocated funds to restore and expand the temple and renamed it as Dajue Temple. In the 59th year of the reign of Emperor Kangxi of the Qing Dynasty (1720 A.D.), large-scale renovations of the temple were conducted. In the 12th year of the reign of Emperor Qianlong (1747 A.D.), the principal halls were repaired.

The temple, which faces east, occupies 9,500 square meters of land, embodying the towards-the-sun convention of temples in the Liao Dynasty. The buildings can be divided into the middle, the eastern and the western axes. On the middle axis, from the front to the rear, are the front gate, the stone bridge, the Hall of Heavenly Kings, the Hall of the Mahavira, the Hall of the Amitayus and the Hall of Dabei (Great Mercy).

功德池
The Gongde Pond

大雄宝殿
The Hall of the Mahavira

三世佛
Buddhas in the Three Periods

天王殿
The Hall of Heavenly Kings

无量寿佛殿匾额
The Board of the Hall of the Amitayus

殿内佛像
Buddhist Statues in the Hall of
the Amitayus

千年银杏树
The Thousand-year-old Ginkgo

鼓楼
The Drum Tower

松抱塔
Pine Tree Embracing Pagoda

辽代清水院碑
Qingshuiyuan Tablet of the Liao Dynasty

塔身砖雕
Brick Carvings on the Body of the Pagoda

# 大慧寺
## Dahui Temple

大慧寺位于北京市海淀区魏公村大慧寺路10号。2001年被公布为全国重点文物保护单位。

大慧寺为明代正德八年(1513年)司礼太监张雄创建，敕名大慧寺。嘉靖年间，因皇帝崇信道教，太监们在寺后建了一座真武庙，以防佛寺被毁。明万历二十年(1592年)、清乾隆二十二年(1757年)两次重修。清末寺院与道院逐渐倾圮，一些殿宇被拆。建国初，尚有照壁、山门、东西配殿等，现只有大慧寺的大悲宝殿留存下来。

来到大慧寺能够同时欣赏到明代的建筑、雕塑、壁画三种艺术杰作。

Located at Weigongcun in Haidian District, the Dahui Temple was listed as a national key relic under special preservation in 2001.

The temple was built by Zhang Xiong, a eunuch, and named Dahuisi in the 8th year of the reign of Emperor Zhengde of the Ming Dynasty (1513 A.D.). During the reign of Emperor Jiajing, who was a devotee of Taoism, eunuch Mai Fu built two Taoist abbeys around the Dahui Temple: Yousheng to the left and Zhenwu behind it. The temple was repaired and reconstructed twice during the Ming and Qing Dynasties. Most of the buildings were gradually dismantled and destroyed in the last Qing Dynasty. Now, only the Hall of Great Mercy remains.

One can enjoy the outstanding architectures, sculptures and murals in Ming Dynasty within one temple of Dahui.

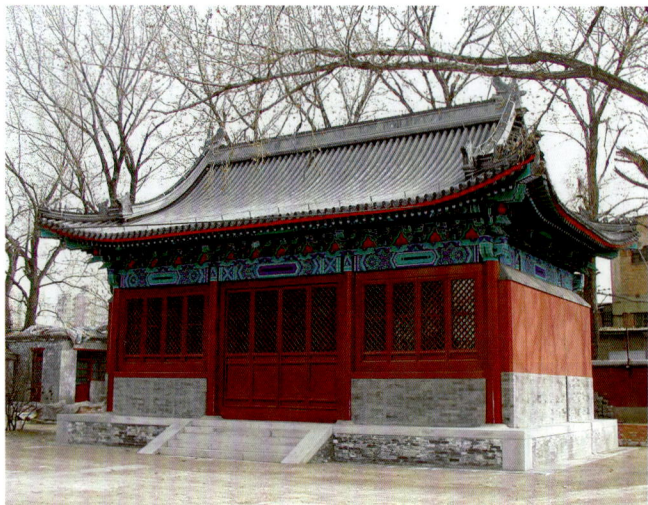

前殿
The Front Hall

大悲宝殿
The Hall of Great Mercy

大悲宝殿是一座面阔五间、进深三间的灰筒瓦重檐庑殿顶的大殿，两层檐之间有采光用的菱花窗，殿内枋柱插头处均安放彩色小佛。明间正面原有一尊高五丈的千手千眼观世音菩萨铜立像，因此人们都把大慧寺叫做大佛寺。铜像在抗日战争时期毁掉，现存的木胎彩绘的两个胁侍是后代补塑的。

壁画
Murals

位于诸天像背后的墙面上，内容是描写一个终生为善者超生得道的故
事，题材新颖、色彩鲜艳、人物描绘细致而传神。

诸天像
Statues of Devas of Law Protection

大悲宝殿内的两面山墙和后檐墙前面伫立28尊明代泥塑彩绘的诸天像。二十八诸天像是佛的护法神，当时的雕塑艺术匠师们，根据诸天神的不同性格，塑造出彼此各异的神态，再加以服饰的衬托和彩色的渲染，更显得形象生动、气势不凡，不愧为现存明代塑像艺术的杰作。

诸天像（局部之一）
Statues of Devas of Law Protection（detail Ⅰ）

# 广化寺
# Guanghua Temple

广化寺位于北京市西城区后海鸭儿胡同31号，是北京著名的佛教十方丛林。1984年被公布为北京市文物保护单位。

广化寺始建于元代，明天顺至成化年间(1457－1487年)予以重修。由于得到内府太监苏诚的资助，重修后的广化寺规模宏大。到明万历二十七年(1599年)广化寺成为净土宗道场。清光绪二十年(1894年)再次重修殿宇。清末民初，广化寺一度成为京师图书馆。

广化寺占地面积20余亩，合13858平方米，建筑共分五路。除一般寺庙的中、东、西路以外，又在两旁增建东二路和西二路，使院落更显得宏阔而严整，殿宇达329间。中路是全寺的主体建筑所在。沿中轴线主要殿堂依次为影壁、山门殿、大雄宝殿、藏经阁，殿两侧对称排列着钟楼、鼓楼、伽蓝殿、祖师殿、首座寮与维那寮。

东路由戒坛、斋堂、学戒堂、引礼寮等殿堂组成。西侧两路各有二进院落，有大悲坛、观音阁、地藏阁、方丈室、法堂、祖堂等。院落之间回廊环绕，僧房毗连，形成一座大四合院中有众多小四合院，即"院中有院"的建筑特色。整座寺庙古柏苍翠，花草溢香，曲径通幽。

Located at No.31 Ya'er Hutong, Shishahai Lake, Xicheng District, Guanghua Temple is titled "the First Temple under Heaven". It was listed as a Beijing's relic under preservation in 1984.

First built in the Yuan Dynasty, reconstruction of the temple began during the reign of Emperor Tianshun of the Ming Dynasty (1457 A.D.) and Su Cheng, an official eunuch, raised the fund. It was completed in 1487 and was renovated again in the 20th year of the reign of Emperor Guangxu of the Qing Dynasty (1894 A.D.). During the last years of the Qing Dynasty and the early years of the Republic of China, the temple used to be the Peking Library.

The temple covering an area of 13,858 square meters and having 329 buildings is divided into five axes. On the middle axis, the main buildings, from the front to the rear, are the screen wall, the front gate, the Hall of Heavenly

山门
The Front Gate

Kings, the Hall of the Mahavira, the two-storied Sutra Hall. The front gate is 3 bays wide with a gable-and-hip roof. The Hall of Heavenly Kings is 3 bays wide with a hip roof. The Hall of the Mahavira is 5 bays wide with a double-eaved gable-and-hip roof. The two-storied Sutra Hall has a flush gable roof.

铜钟
The Bronze Bell

天王殿
The Hall of Heavenly Kings

五佛宝殿
The Five-Buddha Hall

大雄宝殿佛像
Buddhist Statues in the Hall of the Mahavira

藏经楼
The Two-Storied Sutra Hall

法器
The Wooden Fish

伽蓝殿
The Qielan Hall

# 广济寺
# Guangji Temple

广济寺位于北京市西城区阜成门内大街25号，是中国佛教协会所在地。2006年被公布为全国重点文物保护单位。

寺院是在金朝西刘村寺基址上建起，元朝改称"报恩洪济寺"，元末毁于战火。明天顺年间（1457－1464年），山西僧人普慧、圆洪等法师于废址上重建寺庙，历经两年建成。明成化二年（1466年）下诏命名为"弘慈广济寺"。明清两朝三次扩建，其中，清康熙三十八年（1699年）增建御制碑文匾额，并增塑了释迦牟尼等鎏金佛像。1921年毁于火，1924年重建，1934年1月火灾严重，正殿和后殿烧毁，明代经殿和国外进贡白檀释迦牟尼立像俱焚，次年重修。1952年、1972年、2000年又经过三次大规模修建。

广济寺坐北朝南，总占地面积35亩，在中轴线上从山门至后殿共五层殿宇，包括山门、天王殿、大雄殿、圆通殿（观音殿）、藏经阁（舍利阁）等，西院有持律殿、净业堂和云水堂。寺庙的西北隅是建于清康熙三十七年（1698年）至今保存完好的戒坛殿和汉白玉砌成的戒坛，今称"三学堂"。寺中旧有古树一棵，树旁立石碑上刻乾隆帝御制《铁树歌》。

Located on No.25 Fuchengmennei Street, Guangji Temple is now the headquarters of the Chinese Buddhism Association. It was listed as a national key relic under special preservation in 2006.

The temple was originally built in the Jin Dynasty, but was completely destroyed during the chaos of the Jin and Yuan Dynasties. During the reign of Emperor Tianshun

山门
The Front Gate

天王殿
The Hall of Heavenly Kings

of the Ming Dynasty (1457-1464 A.D.), Puhui and Yuanhong, two monks from Shanxi Province, came here and rebuilt this temple at the same place. During the Ming and Qing Dynasties, the temple had undergone three expansions. The temple was again destroyed by fire in 1934. It was rebuilt and renovated again in 1952, 1972 and 2000.

The Guangji Temple, which faces south, covers an area of 35 Mu. It comprises five courtyards one behind another and on the middle axis stands such important buildings as the front gate, the Hall of Heavenly Kings, the Mahavira Hall, the Kwan-yin Hall, the two-storied Sutra Hall etc. from the south to the north.

铜佛
The Bronze Buddhist Statue

大雄宝殿
The Hall of the Mahavira

大雄宝殿是第二进院的正殿，面阔五间，单檐庑殿顶、黄琉璃瓦屋面。殿前有一尊乾隆五十八年(1793年)铸造的铜宝鼎，高约两米，鼎身铸有佛教八宝(轮、螺、伞、盖、花、瓶、鱼、结)等花纹，造型古朴大方、工艺精湛，是珍贵的艺术珍品。

藏经阁(舍利阁)
The Two-Storied Sutra Hall

位于第四进院，是二层的后罩楼，上为舍利阁、绿琉璃瓦顶，下为多宝殿、黄琉璃瓦檐，陈列国际佛教友人所赠珍品，十分珍贵。舍利阁曾于1955－1964年供奉灵光寺"佛牙舍利"，现为藏经阁，珍藏佛教经书10万余册，并藏有房山云居寺石经拓片。藏经阁内还有1721－1753年甘肃临潭县卓尼寺雕版印刷的一部藏文《大藏经》，共231包，是佛教典藏中的珍贵文本。另有宋、明时期血写佛经更为珍贵。

# 觉生寺(大钟寺)
# Juesheng Temple

觉生寺位于北京市海淀区北三环西路北侧，因为寺内珍藏一口明代铸造的大钟而闻名，俗称"大钟寺"。1996年被公布为全国重点文物保护单位。

1733年，清雍正皇帝敕令建此寺院，次年落成。觉生寺建造之初为佛徒、僧人寂静清修和善男信女顶礼朝拜之地；乾隆五十二年(1787年)遇久旱无雨，皇上到此祈雨；该寺作为皇家祈雨活动场所一直延续到清朝末年。

觉生寺坐北朝南，前后五进院落。由南向北依次为照壁(已毁)、山门、钟鼓楼、天王殿、大雄宝殿、后殿、藏经楼、大钟楼与东西翼楼。此外，还有六座配庑分布在两侧。大钟楼内高悬的永乐大钟，是明代永乐年间(15世纪初)，明成祖朱棣迁都北京后下令铸造的，是成祖迁都北京后三大工程之一(另两项为建皇宫、修天坛)，距今已560多年。

Situated in the northern section of the North 3rd Ring Road in Haidian District, the Juesheng Temple is known as the national treasure Yongle Huge Bell being kept in the temple, so it is commonly called the Great Bell Temple. It was listed as a national key relic under special preservation in 1996.

In 1733, Emperor Yongzheng of the Qing Dynasty issued an imperial edict to build the temple, which was completed the follow year. From 1787 to the end of Qing Dynasty, the temple served as a place prayed for rain by imperial family.

The temple, which faces south, comprises five courtyards one behind another. The principal structures, from the south to the north, are the front gate, Drum and

山门
The Front Gate

位于该寺最南面，面阔三间，进深五檩，歇山式调大脊筒瓦屋面，檐下施一斗二升交麻叶墨线斗拱，明间六攒，次间五攒，木构架绘墨线大点金旋子彩画，明间石券门上卧匾"敕建觉生寺"，次间石券窗；山门两侧各有一边门，两边撒山影壁立于石须弥座上，上身中心、四岔角砖雕卷草宝相花纹。门前左右各有一对石狮子。

Bell Towers, the Hall of Heavenly Kings, the Hall of the Mahavira, the rear hall, the two-storied Sutra Hall, the Great Bell Tower and the eastern and western wing buildings. The Yongle Huge Bell hanging in the Great Bell Tower was cast during the Yongle period of the Ming Dynasty, so the bell has a history of more than 560 years.

大钟楼
The Great Bell Tower

大钟楼位于第五进院，高20米，是寺内独具特色的核心建筑，它矗立在青石台基的月台之上，汉白玉栏板望柱，前出双垂带七级台阶。整个钟楼上圆下方，象征"天圆地方"。钟楼下层为正方形，面阔三间，明间抹隔扇门、窗，其上卧匾"华严觉海"，门前双垂带五级台阶，次间四抹隔扇门、窗；上层为八角攒尖筒瓦屋面，檐下施一斗二升交麻叶墨线斗拱，雅五墨彩画。12根柱子将圆楼分为十二间，每间上有斗拱四攒，下有套方窗四扇，再下为如意挂檐板。楼内悬挂大钟，内设旋梯可供上下。青石台基上砌有八角形"散音"池，池深70厘米，直径4米，池口距钟口1米，钟响时，有很好的共鸣作用。

永乐大钟
The Yongle Huge Bell

永乐大钟（局部）
Section of the Yongle Huge Bell

# 法海寺
# Fahai Temple

法海寺位于北京市石景山区模式口翠微山南麓。1988年被公布为全国重点文物保护单位。

据寺内碑文记载，此寺为明正统四年(1439年)太监李童等集资，由工部营缮司建造的。四年后建成，由英宗赐名"法海禅寺"。明清时曾多次大修。1953年重修一次。1982年又将山门、大雄宝殿和东西庑殿重修。1985年重修天王殿。1995年重修大雄宝殿。2005年复建藏经阁。

法海寺占地100500平方米，现仅存大雄宝殿、山门数处。大雄宝殿面阔五间，单檐庑殿顶，与智化寺、隆福寺诸殿同为明代前期按官式建造的重要佛殿。法海寺大雄宝殿较其他各寺更有价值之处是殿内明代壁画尚存。此殿壁画是北京地区现存最精美、最巨大的壁画，从佛寺由工部监造的情况看，画工也应当是当时名手，它在国内现存明代壁画中亦属上品，在我国壁画史上也占有极高的地位。

山门
The Front Gate

大雄宝殿
The Hall of the Mahavira

Located at the southern foot of Cuiwei Mountain in Shijingshan District, it was listed as a national key relic under special preservation in 1988.

According to the record, construction of the temple was begun in the 4th year of the reign of Emperor Zhengtong of the Ming Dynasty (1439 A.D.) by the Board of Works. Funds were raised by Li Tong, a eunuch. The temple was completed four years later and named Fahai Chansi by Emperor Yingzong. Then, the temple was renovated several times during the Ming and Qing Dynasties.

The Fahai Temple occupies an area of 100,500 square meters, now, only the Mahavira Hall and the front gate are preserved. The hall is 5 bays wide with a hip roof and is decorated with remarkable Ming Dynasty murals which are the most exquisite and largest murals in Beijing. According to historical records, the paintings are executed by famous artisans. The murals of the Fahai Temple hold an important status in the mural history of China.

法海寺壁画
Murals of the Fahai Temple

位于大雄宝殿的佛像后扇面墙的背后，正对后门的地方，绘有观音（中）、普贤(左)、文殊(右)三大士坐像，后壁两侧有礼佛图两幅，东西山墙于罗汉塑像后绘有佛、菩萨像。其中以三大士像最为精彩，面貌饱满圆润、服饰华丽、衣纹流畅生动，尚有元代遗韵。

法海寺壁画（局部）
Section of Murals of the Fahai Temple

法海寺壁画（局部）
Section of Murals of the Fahai Temple

# 灵岳寺
# Lingyue Temple

灵岳寺位于北京市门头沟区斋堂镇北部5公里的白铁山上，是一座大型佛教寺院。2003年被公布为北京市文物保护单位。

灵岳寺创建于唐贞观年间(627－649年)。辽代重建，称白贴山院，金代改称灵岳寺。元至元三十年(1293年)和至正年间均有整修，清康熙二十二年(1683年)、雍正十一年(1733年)两次整修。

寺院处于白铁山主峰前的平台上，坐北朝南，中轴线上有山门、天王殿和释迦佛殿。寺南部山门两侧为钟鼓楼遗址。天王殿面阔三间，为悬山式建筑，殿内旧供奉四大天王、韦驮、接引佛塑像。释迦佛殿面阔进深均为五间，单檐庑殿顶，面积达100余平方米，檐下重昂五彩斗拱，拱眼壁处彩绘佛像，殿堂虽经历代修缮，仍保存元代遗韵。殿内旧供奉一佛二菩萨像，柳木雕刻，高近4米，释迦牟尼端坐莲台之上，两侧有阿难、迦叶立像，雕刻传神，可惜于1954年拆毁。寺内现存至元三十年(1293年)《重修灵岳寺记》碑、清康熙二十二年(1683年)《重修灵岳禅林碑记》。

Situated in Baitie Hill, about 5 kilometers to the north of Zhaitang Township in Mentougou District, the Lingyue Temple is a great Buddhist temple. It was listed as a Beijing's relic under preservation in 2003.

The temple was first built during the Zhenguan reign of the Tang Dynasty (627-649 A.D.). It was reconstructed and named Baitieshanyuan in the Liao Dynasty, which

山门
The Front Gate

灵岳寺全貌
A Panorama of the Lingyue Temple

was renamed Lingyuesi in the Jin Dynasty. It underwent two periods of reconstruction in the 22nd year of the reign of Emperor Kangxi (1689 A.D.) and in the 11th year of the reign of Emperor Yongzheng of the Qing Dynasty (1733 A.D.).

The temple, which faces south, is situated on a platform in front of the highest peak in Baitie Hill. On the middle axis are the front gate, the Hall of Heavenly Kings and the Sakyamuni Buddha Hall. The Hall of Heavenly Kings is 3 bays wide with an overhanging gable roof. The Sakyamuni Buddha Hall is 5 bays wide and 5 bays deep with a hip roof, covering an area of more than 100 square meters.

释迦佛大殿
The Sakyamuni Buddha Hall

释迦佛大殿内彩画
Coloured Paintings in the Sakyamuni Buddha Hall

# 承恩寺
## Cheng'en Temple

　　承恩寺位于北京市石景山区模式口大街东口路北，其西北700米是明代的法海寺，西600米有明代太监田义墓。2006年被公布为全国重点文物保护单位。

　　承恩寺坐北朝南，四进院落，周身以院墙围护，占地约30亩。由南向北依次为山门、天王殿、大雄宝殿、法堂，为典型的寺院格局。较为独特之处是寺院院墙四角设有瞭望碉楼，寺内还有地道和地下室与碉楼相连，为明清寺庙所罕见，在北京的庙宇中亦绝无仅有的。相传承恩寺为明代著名太监刘瑾所建，因其不满足"九千岁"之称，承恩寺便是他演兵造反的大本营。此说虽不见经传，但从寺院的格局分析，其军事目的是显而易见的。

山门及碉楼
The Front Gate and the Watchtower

山门三间，单檐歇山顶，筒瓦屋面，汉白玉券门上石额有曰："敕建承恩禅寺"。山门东、西两侧各有一便门。山门以北为天王殿，单檐歇山顶，筒瓦屋面，面阔三间，檐下施一斗二升三蝠云斗拱，木构架绘旋子彩画。

The Cheng'en Temple is located at the north of the eastern end of Moshikou Street in Shijingshan District, about 700 meters to the southeast of Fahai Temple and 600 meters to the east of Tian Yi's Tomb.It was listed as a national key relic under special preservation in 2006.

The temple covering an area of about 30 Mu, which faces south, comprises four courtyards one behind another and is surrounded by walls. The principal structures, from the south to the north, are the front gate, the Hall of Heavenly Kings and the Hall of the Mahavira. There are four watchtowers on the four corners of the wall, which are connected by subways and basements, this is an unique feature among temples in Ming and Qing Dynasties. It is said that Liu Jin, a famous eunuch of the Ming court, built the temple with some military purpose.

明代彩画
Coloured Paintings of the Ming Dynasty

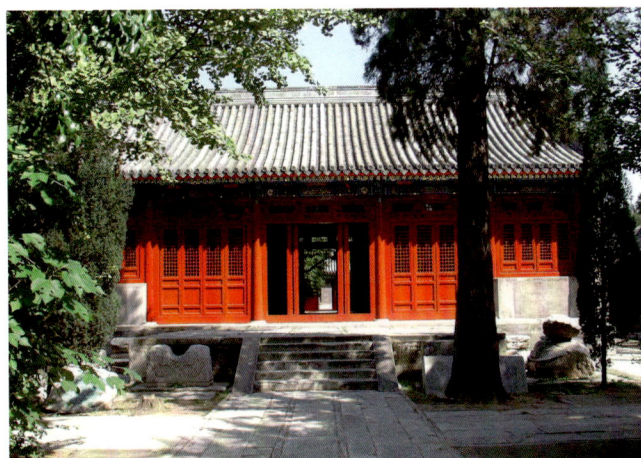

大雄宝殿
The Hall of the Mahavira

# 普度寺
# Pudu Temple

普度寺位于北京市东城区南池子大街内普庆前巷，俗称玛哈噶喇庙，是一座喇嘛庙。1984年普度寺大殿被公布为北京市文物保护单位。

明代时该寺是皇城东苑洪庆宫（又名南城，或小南城）的一部分，清朝入关后，此地为多尔衮的王府（摄政王府）。清康熙三十三年（1694年），改建成玛哈噶喇庙（玛哈噶喇为梵语，意为大黑神）。清乾隆四十年（1775年）又进行了一次修缮，次年赐名普度寺。

现在寺院由于拆毁较多，古建筑仅存山门、大殿及一座朵殿。普度寺整体建筑都建在一座砖石砌筑的高台上，山门三间，硬山调大脊，绿琉璃筒瓦屋面，拱券门。普度寺大殿建造在呈"凸"形汉白玉石须弥座上，座之比例及雕刻均有明代特征。殿面阔七间，进深三间，全殿外加周围廊。单檐歇山顶，削割筒瓦绿琉璃瓦剪边，前出抱厦三间，歇山卷棚顶，绿琉璃瓦黄剪边。大殿周围绕以36根檐柱，柱间带雀替，山墙与檐墙用大青砖平铺顺砌，墙下部用绿色六边形琉璃釉砖拼成几何图案，大殿窗棂低矮，殿内东部隔出二间内室。外檐出檐为三层椽，无斗拱，在柱头装饰兽面木雕，雀替形式特殊，室内彩画还有不少博古题材，整个大殿显得宏伟壮观，建筑具有明显的关外满族宫室特征，具有如此鲜明满族风格的古建筑在北京仅此一处。大殿之西尚存方丈院北房五间，北房两侧连接一段游廊。

Located at Pudusi Qianxiang Nanchizi Street in Dongcheng District, the Pudu Temple is a Lama temple, commonly known as Mahagala Temple. It was listed as a Beijing's relic under preservation in 1984.

Shortly after the Qing Dynasty was founded, it served as Duo'ergun's Residence. In the 33rd year of the reign of Emperor Kangxi of the Qing Dynasty (1694 A.D.), the residence was rebuilt as the Mahagala Temple and was renovated and renamed Pudu Temple in the 40th year of the reign of Emperor Qianlong (1775 A.D.).

The whole temple was built on a platform made of bricks. The front gate is 3 bays wide with a flush gable roof and ridge. The roof is covered with green glazed tube-shaped tiles. The hall of Pudu Temple is situated on a white marble stone terrace shaped like "凸", the scale and carving of which can be dated back to the Ming Dynasty. The hall is 7 bays wide and 3 bays deep with a gable-and-hip roof and all enclosed with corridors. The roof is covered with tube-shaped tiles with a green edge. The hall has 3 bays on its south with a gable-and-hip and rolled pitched roof. The roof is covered with green glazed tiles with a yellow ridge.

大殿
The Hall of Pudu Temple

大殿(局部)
Section of the Hall of Pudu Temple

普度寺石雕
Stone Carvings on the Hall of Pudu Temple

大殿彩画
Coloured Paintings of the Hall of Pudu Temple

# 智化寺
# Zhihua Temple

智化寺位于北京市东城区禄米仓街5号，是北京现存最大的明代建筑群之一。1961年被公布为全国重点文物保护单位。

据该寺《敕赐智化禅寺之记》及《敕赐智化禅寺报恩碑》记载，智化寺始建于明正统九年(1444年)正月初九，完成于同年三月初一。原为司礼太监王振的家庙。清末及民国，寺内文物不断散失，建筑亦多失修。解放后，人民政府为保护这座珍贵的明代建筑，加强了管理并不断进行维修；1987年，又对智化寺各殿进行修缮。

智化寺坐北朝南，共有五进院落。山门面阔三间，进深一间，为砖砌仿木结构，前檐石额："敕赐智化寺"。山门对面原有砖砌影壁，建国后拆除。第一进院内有钟鼓楼、智化门(今已无存)等。第二进院内有智化殿及东西配殿等，智化殿在智化门正北，坐北朝南。殿内中央有汉白玉须弥座，上供佛像，左右两侧各列罗汉10尊，北侧亦有佛座，现均不存。第三进院内有如来殿(万佛阁)，殿上下两层，墙壁遍饰佛龛，原置小像约9000躯(现缺损很多)，故上檐榜书"万佛阁"。明间中央藻井作斗八式杂饰云龙，于20世纪30年代初为寺僧盗卖，现存美国纳尔逊博物馆。第四进院内有大悲堂(后殿)等。第五进院内有万法堂等。又有西跨院为方丈院，在大悲堂西侧；东跨院是后庙，在大悲堂东侧。智化寺建筑，与唐宋相比，已发生了变化，与清代相比，又有所不同。总体说来，在规范方面仍受宋《营造法式》的一些影响，但在局部结构与装修方面有所变化。

山门
The Front Gate

智化殿
The Zhihua Hall

Located at 5 Lumicang Hutong in Dongcheng District of Beijing, the Zhihua Temple is one of the largest complex of temple buildings of the Ming Dynasty in Beijing. It was listed as a national key relic under special preservation in 1961.

First built in the 9th year of the reign of Emperor Zhengtong of the Ming Dynasty (1444 A.D.), the temple was the family shrine of Wang Zhen, a eunuch of the court.

The temple Facing south, it comprises five courtyards one behind another. The front gate is 3 bays wide and 1 bay deep, on the opposite side of which was a screen wall. In the first courtyard are the Zhihua Gate and Drum and Bell Towers. In the second courtyard is the Zhihua Hall, on the eastern and western sides of which there are wing halls. In the third courtyard there is the Hall of the Tathagata, which has two storeys and enshrines the statue of the Tathagata. On the upper floor, there are more than 9,000 little niches in the walls, and therefore it is also called Pavilion of Ten-Thousand Buddha. On the top of the bright room inside the pavilion, there was an exquisitely carved sunk panel. However from 1930 to 1934, some monks stole and sold them, which are now preserved in the Nelson Museum in United States.

大智殿
The Dazhi Hall

佛像
Buddhist Statue

如来殿(万佛阁)
The Hall of the Tathagata

经幢
The Turning Wheel

万佛造像
Ten-Thousand Buddhist Statues

# 天宁寺塔
# The Pagoda at the Temple of Celestial Tranquility

天宁寺塔位于北京市宣武区广安门外天宁寺东里2号，是辽金北京城中现存的唯一一座完整的建筑物。1988年被公布为全国重点文物保护单位。

天宁寺是北京地区年代最早的庙宇之一，始建于北魏年间，辽陪都时期，在寺后建佛塔，即天宁寺塔，是辽塔中具有代表性的佳作，距今已有900多年历史。寺庙在元末毁于兵火，所有建筑荡然无存，只余下高塔茕子无依。后经历代多次重修，至今保存完好，是研究辽南京城地理位置的重要依据，也是现在北京最珍贵的建筑艺术遗物之一。

天宁寺塔为八角十三层密檐式实心砖塔，通高57.8米，最下面方形平台，平台之上是两层八角形基座，基座上承须弥座。下层须弥座各面以短柱隔成六座壸门形小龛，龛内雕有狮兽头，龛与龛之间雕缠枝莲图案，转角处雕有造型极为生动的金刚力士像。上层须弥座稍小，每面也以短柱隔成五座壸门形小龛，龛内雕坐佛一尊，龛与龛之间雕有守护佛像的金刚力士像，转角处亦雕有顶抗塔基形象的金刚力士像。平座位于须弥座之上，平座每面都有砖雕仿木重拱偷心造斗拱，补间三朵，平座勾栏上雕缠枝莲、宝相花等纹饰。平座之上用三层仰莲座承托塔身。

塔身平面呈八角形，四正面辟拱门，四斜面饰直棂窗，门窗上部和两侧高浮雕金刚力士、菩萨、天部等神像。塔身转角处的砖柱上浮雕升降龙。所有雕饰造型均生动完美，线条流畅，堪称中国古代雕刻艺术的精品。

塔身之上建造十三层塔檐，檐下均施仿木结构的砖制双抄斗拱，第一层为辽代建筑中特有的45°斜拱，其上各层无斜拱使用。各层塔檐自下而上逐层递减，轮廓线形成丰满柔和的收分，各层塔檐的角梁均为木制，各种瓦件和脊兽、套兽等构件，全部用琉璃烧制，建筑工艺十分讲究。塔檐每层均系缀风铃,每逢风起,便发出清脆悦耳的铃声。塔刹为两层八角形仰莲座之上置小须弥座承托宝珠。天宁寺塔的造型雄伟壮丽，稳中挺拔，是辽代佛塔中不可多得的珍贵文物。

Located outside Guang'anmenwai in Xuanwu District, standing in the Temple of Celestial Tranquility, the Pagoda at the Temple of Celestial Tranquility is the oldest, the largest and the highest among ancient buildings still in Beijing. It was listed as a national key relic under special preservation in 1988.

Regarded as one of the most ancient temples in downtown Beijing, the temple was first built during the Northern Wei Dynasty. The pagoda of Celestial Tranquility was constructed in the Liao Dynasty, which has a history of more than 900 years. During the last years of the Yuan Dynasty, the temple suffered from the ravages of war and collapsed. Then, the only extant pagoda had been renovated many times.

Built of bricks the octagonal multi-eaved pagoda with 13 tiers is exactly 57.8 meters high. The pagoda rests on a large square platform, on which is a 2-story octagonal base, on which is a shu-mi-tso ornamented with a single band of relief-carved arched niches. Guardian Kings are carved at each corner. A p'ing-tso is raised on the shu-mi-tso, carved with patterns of lotus scroll.

The four front walls of the octagonal body have arches and on each of the four adjacent walls is a window with vertical bars, which are decorated with Guardian Kings and Bodhisattvas. Above this, the 13 levels rise in a slightly bowed profile. Bells hanging from each story tinkle pleasantly in the wind. The uppermost level is surmounted by a pearl-shaped symbol, which represents the Buddhist faith.

天宁寺塔
The Pagoda at the Temple of
Celestial Tranquility

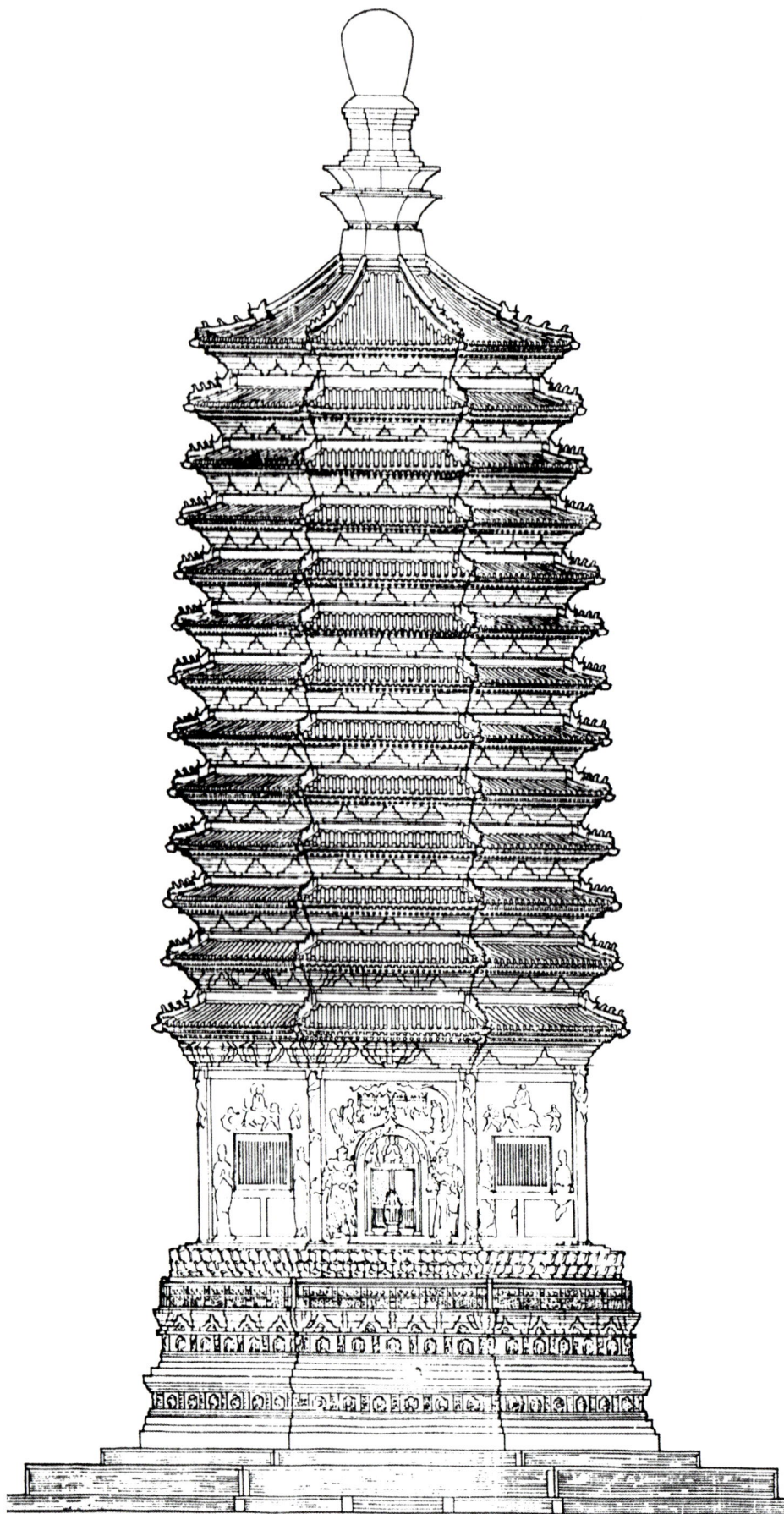

天宁寺塔线图
The Plan of the Pagoda at
the Temple of Celestial
Tranquility

天宁寺塔砖雕
Brick Carvings of the Pagoda at the Temple of Celestial Tranquility

# 真觉寺金刚宝座塔 (五塔寺塔)

# The Diamond Throne Pagoda at the Temple of True Awakening

真觉寺金刚宝座塔位于北京市海淀区西直门外白石桥，是北京现存年代最早、最精美的金刚宝座塔，因其形式是由五座佛塔组成，俗称五塔寺塔。1961年被公布为全国重点文物保护单位。

明永乐年间，印度高僧班迪达向明成祖朱棣进献了五尊金佛和一幅金刚宝座塔的建筑图样，被永乐皇帝封为国师，并下诏依此图形建造佛塔。明成化九年(1473年)，金刚宝座塔竣工，赐名真觉寺。清乾隆年间重修，改为大正觉寺。清末，八国联军侵入北京，寺院建筑荡然无存，唯金刚宝座塔因全部是石质结构而幸存。

金刚宝座塔由宝座和石塔两部分组成。建筑平面呈长方形，南北长18.6米，东西宽15.73米，通高15.7米，其中宝座高为7.7米。宝座内部用砖砌筑，外部全部用青白石包砌。

宝座最下面是须弥座式石基座，上面为座身，分为五层，每层均有挑出的石短檐，檐头刻出筒瓦、滴水等建筑构件。宝座上雕刻着上下五层佛龛，每个佛龛内刻有佛像一尊，形态各异，号称千佛。宝座南北辟券门，内设过室、塔室、塔内中心柱、佛龛、佛像等。拱门券面上刻有狮、象、孔雀、金翅鸟等图像纹饰。南面券门上嵌有"敕建金刚宝座塔，大明成化九年十一月初二日造"石额。金刚宝座塔过室东西两侧辟小券门，各有石阶梯44级，盘旋而上，通向宝座平台。

平台四周建有石护栏围绕，平台上分别建有五座四角密檐式小石塔。石塔平面呈方形，中央的一座高8米、十三层密檐，塔顶是铜质的覆钵式塔刹，传说印度高僧带来的五尊金佛就藏在这座塔中。四隅的小塔高7米，十一层密檐。五座小塔的雕刻精美绝伦，塔身雕饰以五方佛坐像为主要内容，共计佛像1561尊，造型端庄祥和。另外，佛塔还雕刻了大量梵文、藏文，更加突出了藏传佛教的建筑文化。

The Diamond Throne Pagoda at the Temple of True Awakening is situated at Xizhimenwai in Haidian District. Among the pagodas of the same style still standing in Beijng, this is the oldest and the most exquisite. It is popularly referred to as the Five-Pagoda Temple's pagodas, because there are five pagodas standing on a large square foundation know as the throne. The Diamond Throne Pagoda was listed as a national key relic under special preservation in 1961.

This architectural form was introduced to China by an Indian monk during the reign of Emperor Yongle of the Ming Dynasty, and the Diamond Throne Pagoda was completed and named the Temple of True Awakening in the 9th year of the reign of Emperor Chenghua of the Ming Dynasty (1473 A.D.). During the reign of Emperor Qianlong of the Qing Dynasty, the temple was reconstructed and renamed the Temple of Great Righteous Awakening. In the late Qing Dynasty it was looted and burned to the ground by the Eight-Power Allied Force. Today, the only extant relic is the Diamond Throne Pagoda.

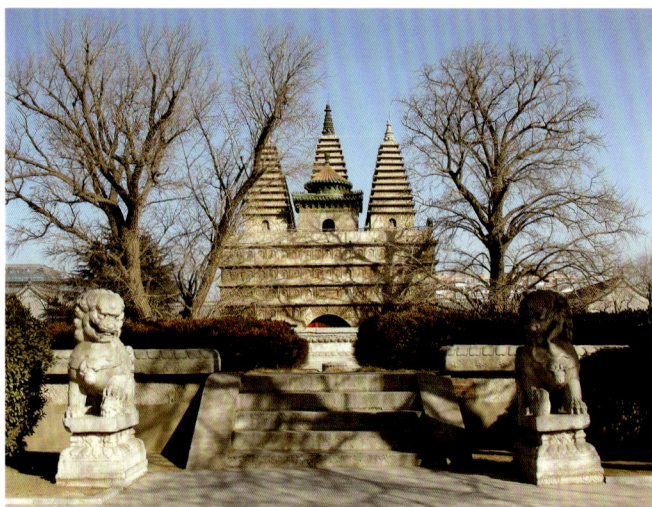

金刚宝座塔
The Diamond Throne Pagoda

金刚宝座顶部小塔
Sub-Pagodas of the Diamond
Throne Pagoda

The pagoda consists of the diamond throne and the stone pagoda, presenting a pattern of rectangle. Standing 15.70 meters high, it measures 18.60 meters from the north to the south and 15.73 meters from the east to the west, the diamond throne of which is 7.7 meters high. It is built with bricks inside and covered with slabs of blue and white granite outside.

Raised on a shu-mi-tso, the diamond throne is divided into five stories and treated with continuous rows of cornices. The four walls of the throne are carved with rows of Buddhas (the One Thousand Sagacious Buddhas). There are arches on the north and south of the throne, on which are carved patterns of lions, elephants, peacocks, redpolls and so on. An arch at the throne on each of the eastern and western sides opens into an inner spiral 44-step stairway that leads to the top of the throne.

The five pagodas rise from their rectangular bases on top of this throne, one in each of the four corners and the fifth in the center. The central pagoda, totaling 8 meters high, is slightly higher than the others, with 13 eaves, two more than those in the corners. The five pagodas are carved with images of Buddha in five directions, as well as carved in Sanskrit and in Tibetan.

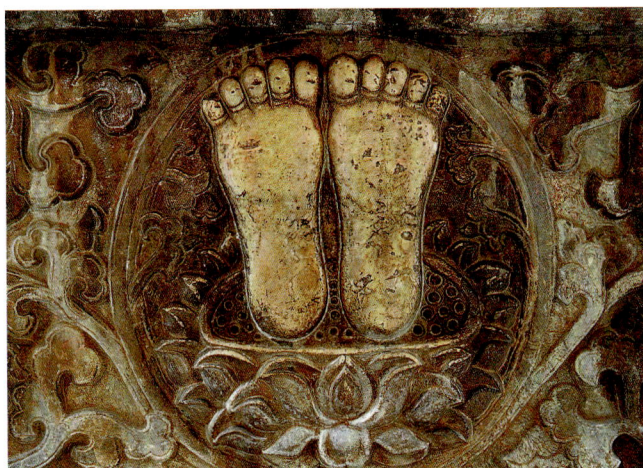

金刚宝座塔主塔——佛足浮雕
Buddha Feet Relieved on the Main Sub-Pagoda of the Diamond Throne Pagoda

中央石塔的塔座南面正中，刻有佛足一双，寓意表示佛迹遍天下，成为金刚宝座塔的重要景观。

金刚宝座塔佛像、菩萨、菩提树
Buddhas, Bodhisattvas and Pipals Carved on a Sub-Pagoda of the Diamond Throne Pagoda

135

# 银山塔林
# Silver Mountain Pagoda Forest

银山塔林位于北京市昌平区下庄乡银山南麓古延寿寺遗址上，是京城北部著名的文物建筑风景区。1988年被公布为全国重点文物保护单位。

银山位于昌平区北部，古时曾是佛家讲经说法的佛教圣地，同时又是文人墨客隐居的最佳场所。唐代高僧邓隐峰在此筑寺修行，银山名声大噪，与江南的镇江金山寺齐名，有"南金北银"之说。辽金时期，寺院建筑群规模进一步扩大，传说建有72座寺庙，僧侣众多，圆寂后均在此修建灵塔。经过数百年的积累，整个银山地区，墓塔林立错落，数不胜数，故当地有"银山宝塔数不清"之说。元明清三代，寺院建筑仍旧不断增加与改建，逐渐成为殿宇与佛塔融为一体的建筑群体。抗战期间，日本侵略军将全部寺院建筑焚毁，烧山毁林，致使原来大批的景观不复存在，只有几座大塔和十余座小塔留存于世。1992年始，十三陵特区对银山塔林及附近古迹进行了发掘和修缮，使这片佛教圣地得以较好的保存。

银山塔林现存17座墓塔，建于金元明清时期，大小高矮不等，高者近20米，直径达7米左右，矮者只有两三米高。从塔的形制上看，大体上可分为两大类：一为密檐式塔，这里保存的几座大型墓塔全部是密檐式结构，是北京地区保存最好、最为集中的密檐式塔群。另一类是覆钵式塔，这里保存的覆钵式塔大部分是石塔，而砖塔的数量则较少。除这两类墓塔形式外，还有一种特殊形式的墓塔，是将密檐和覆钵两种形式的塔结合起来，集于一身，这种建造奇特的异形塔，堪为塔林中的上乘佳作。现在，这些佛塔保存完好，是研究我国古代佛教和砖石建筑的宝贵遗产。

Situated at Xiazhuang Township in Changping District, the Silver Mountain Pagoda Forest is the most notable beauty spots of ancient buildings in the northern part of Beijing. It was listed as a national key relic under special preservation in 1988.

Situated at the north of Changping District, the Silver Mountain used to be the Buddhism Holy Land and the best reclusive spot for literators. During the Liao and Jin

银山塔林
Silver Mountain Pagoda Forest

Dynasties the temple complex was further enlarged and it is said that at that time there were 72 Buddhist temples. During the Yuan, Ming and Qing Dynasties, the temple complex was continuously enlarged and rebuilt to become a group of buildings in combination of halls and pagodas. However, during the War of Resistance Against Japan, all the temple buildings were burned down by Japanese invaders and only the pagodas remained. From 1992, the Silver Mountain Pagoda Forest and near remains had been unearthed and renovated by the Shisanling Special District.

On the Silver Mountain Pagoda Forest stand 17 pagodas. The highest one is nearly 20 meters, while the lowest is only 2 or 3 meters. According the types of the pagoda, they are divided into Multi-eaved pagodas and pagodas of upside-down alms bowl.

# 万佛堂、孔水洞石刻及塔
# Wanfotang and Kongshui Cave

万佛堂、孔水洞石刻及塔位于北京市房山区河北镇万佛堂村西约200米处。2001年被公布为全国重点文物保护单位。

万佛堂、孔水洞石刻及塔坐落于龙泉寺内，该寺创建于唐代宗大历五年(770年)。万佛堂塔有两座，均位于孔水洞的左右两侧，分别建于元代和辽代。其中洞后翼山岗上的坨里花塔，是北京现存最漂亮的辽代花塔。花塔又称华塔，多盛行于中国北方一带，现存尚有七八座。"花塔"一词由来可能是因为这类塔体十分肥硕，雕饰甚多，犹如一根盛开的花棒。花塔雕饰题材以佛学为主，并具有一定的佛教内涵，即表现毗卢舍那佛所居住的"莲花藏世界"。传说在该世界内毗卢舍那佛可化身为1000个释迦，分居在大千世界。花塔塔体的每一座佛龛，即代表毗卢舍那佛的一个化身，共同构成莲花藏世界。据此佛教含义，因而称这种表现莲花千佛造型的塔为华塔。可以说华塔造型是佛塔建筑开始摆脱仿木构形式的束缚，努力追求表现宗教意识，从而开创了佛塔艺术创新的道路。万佛堂塔高20米，在花束形的圆形塔体上，层层缠绕着屋宇形式的佛龛，龛内雕有座佛一尊，龛下托以圆雕的狮、象等形式的龛座，上下层龛位互相叠错，具有渐变的、有韵律的图案组合。塔体雕刻内容丰富，表现手法细腻。该塔外观造型优美，雄浑壮丽。花塔上有"咸雍六年(1070年)"、"寿昌七年(1101年)"等题记，为研究中国花塔出现的年代提供了实物资料。

Located at the west of Wanfotang Village in Hebei Township of Fangshan District, the Wanfotang and Kongshui Cave was listed as national key relic under preservation in 2001.

The Wanfotang and Kongshui Cave are situated in the Longquan Temple which was first built in the 5th Dali Year of the Daizong reign of the Tang Dynasty(770 A.D.). Standing on both sides of the Kongshui Cave the two pagodas were built respectively in the Yuan and Liao Dynasties, among which the Tuolihua Pagoda standing on the hill behind the cave is the most fascinating flowery pagoda of Beijing built in the Liao Dynasty. Flowery pagoda, also named Hua pagoda, prevailed in North China with only seven to eight being left currently. The word *flowery pagoda* comes from the huge body and flourishing sculptures of such type of pagoda, resembling a blooming flower stem. On the theme of Buddhism with Buddha as connotation, the sculptures embody the World Hidden in the Lotus the Vairocana Buddha resides in. It is said that in such a world the Vairocana Buddha can assume a thousand incarnations of the Sakyamuni. Each Buddha niche in the pagoda represents an incarnation of the Vairocana Buddha and all of them constitute the World hidden in the Lotus. The pagoda is 20 meters high and roof-style niches interweaves layer by layer along the bouquet-like round body with a Buddha sitting in each niche. The niches are supported by bases inscribed with lions, elephants, etc. and the niches are overlapped forming a pattern combination of gradual change and rhyme. The contents of sculptures on the body are rich, manifested by exquisite method.

万佛堂
The Hall of Ten-Thousand Buddhas

花塔
The Flowery Pagoda

唐代"万佛法会图"浮雕
Carvings in Relief of the Tang Dynasty

# 清净化城塔
## Qingjing Huacheng Pagoda

清净化城塔位于北京市朝阳区安定门外西黄寺大街中路路北，是北京现存的金刚宝座塔中与印度金刚宝座塔形式最相近的一种建筑类型。2001年被公布为全国重点文物保护单位。

西黄寺是西藏活佛达赖和班禅来京的驻锡之所，初名达赖庙，是清王朝为西藏宗教领袖达赖五世修建的，始建于清世祖顺治九年(1652年)。1780年，班禅六世来京谨见乾隆皇帝后因病在西黄寺圆寂，为了超度班禅，敕建六世班禅衣冠塔，赐名"清净化城塔院"以示纪念。咸丰十年(1860年)，英法联军侵入北京，寺院被毁。光绪三十四年(1908年)，重修西黄寺。民国时期，寺院损坏严重，只剩下清净化城塔院等部分建筑保存下来。新中国成立后，政府拨款对清净化城塔及清净化城塔院进行了大规模修缮。

清净化城塔是西黄寺的标志性建筑之一。整个塔形为印度佛陀伽耶式，即大塔四角建小塔。五塔全部建在3米多高的汉白玉石台基上，全塔除塔顶外，全部用白石砌成，周围有栏杆、牌坊和辟邪。四角各有高7米塔式经幢1座，经幢共分五层，每层供有八座佛像，幢身下层绘刻经文。中央为主塔，主塔高约15米，其形制为藏式，基座呈八角形，须弥座各层均饰以卷草、莲瓣、云彩、蝙蝠等花纹，座的八面各雕佛传奇画一幅，图案雕刻神像、罗汉、信徒，并衬以殿宇、树木等，内容丰富，雕刻艺术精湛。须弥座转角处雕力士像一尊，赤背赤足，孔武有力。须弥座上承托覆体式塔身，正面佛龛内有浮雕三世佛，两侧雕菩萨。塔顶为铜制鎏金双层莲花、相轮、宝瓶盖顶。

清净化城塔用藏传佛教的塔式为主体，形成了融合汉族、藏族和印度佛教诸风格于一体的艺术性建筑，气势宏伟，雕刻精美，享有"北京白塔之冠"的美誉。清净化城塔是清代佛塔建筑艺术的杰作，具有极高的文物和艺术价值。

Located in the Xihuang Temple outside Andingmen in Chaoyang District, the Qingjing Huacheng Pagoda is an architectural type resembling mostly the Indian diamond throne pagoda. It was listed as a national key relic under special preservation in 2001.

As the residence for Dalailama and Banchan in Beijing, the Xihuang Temple, originally named Dalailama Temple, was built by the Qing Dynasty for Dalailama 5th in the 9th year of the reign of Emperor Shunzhi(1650 A. D.). In 1780, Dalailama 6th paid a visit to Emperor Qianlong and afterward fell ill and finally died in the Xihuang Temple. In order to release his soul, a pagoda storing his clothes was built entitled Qingjing Huacheng Pagoda in his honor. In the 10th year of the reign of Emperor Xianfeng (1860 A.D.), Anglo-French Allied Forces invaded Beijing and destroyed the temple. During the Republic of China, the temple was damaged seriously with partial buildings including the Qingjing Huacheng Pagoda being left. After the foundation of the People's

清净化城塔前石牌坊
The Stone Archway in Front of the Qingjing Huacheng Pagoda

Republic of China, the government appropriated special funds for large-scale repairs of the Qingjing Huacheng Pagoda and the courtyard.

Small pagodas were built on the four corners of big pagoda. The five pagodas were all built on the 3-meter-high white marble base and the bodies are made up of white stones except for the tops. 7-meter-high tower-like pillars in five tiers were built on each corner and in each tier set 8 niches. The bottom of pillars is carved with Buddhism scripts. The 15-meter-high main pagoda situates in the middle and is built in Tibetan style, with octagonal base and the shu-mi-tso is decorated with patterns of curling grass, lotus petal, cloud, bat etc. Buddhist legend drawings are carved exquisitely on sides of the shu-mi-tso with rich contents encompassing arhats, disciples of the Buddha etc.

清净化城塔
The Qingjing Huacheng Pagoda

清净化城塔主塔覆钵浮雕
Carvings in Relief on the Upside-Down Alms Bowl of the Main Sub-
Pagoda of the Qingjing Huacheng Pagoda

清净化城塔主塔雕——释迦牟尼成道图
Carvings in Relief on the Main Sub-Pagoda of the Qingjing Huacheng
Pagoda

清净化城塔主塔石雕——力士
Carvings in Relief on the Main Sub-Pagoda of the Qingjing Huacheng Pagoda

# 慈寿寺塔
## The Pagoda at the Temple of Benevolence and Longevity

慈寿寺塔位于北京市海淀区阜外八里庄北里3号，是北京最高、最壮观的佛塔之一。1957年被公布为北京市文物保护单位。

慈寿寺塔原名永安万寿塔，因建于慈寿寺内，故俗称慈寿寺塔，当地人又称玲珑塔。该寺始建于明万历四年(1576年)，敕名慈寿寺。清乾隆二十二年(1757年)时，曾下旨修葺该寺。清光绪年间，寺毁于火灾，仅存万寿塔和塔前的两座石碑。

慈寿寺塔平面呈八角形，为十三层密檐式实心砖塔，通高50米，没有阶梯，不能攀登。整座塔分塔基、塔身和塔刹三部分。塔基又分上下两部分，下为条石砌筑的平台，上是双层的须弥座，须弥座上有四十座壸门形小龛，龛内雕有200多个人像，神态各异，栩栩如生，每个人像都是一个教化人类敬佛向善的佛教故事。须弥座上部刻有笙、箫、琴、瑟等全行乐器图案，这在佛塔中是极为罕见的。须弥座上以三层仰莲瓣承托塔身。塔身平面呈八角形，四正面辟拱门，四斜面饰券窗，门窗装修为仿木构形式，门窗上部和两侧高浮雕金刚力士、菩萨、天部等神像。塔身转角处的砖柱上浮雕升降龙。所有雕饰造型均生动完美、线条流畅，是中国古代不可多得的雕刻艺术的珍品。

Situated at No.3 Balizhuang in Haidian District, the Pagoda at the Temple of Benevolence and Longevity is one of the highest and most splendid pagodas in Beijing. It was listed as a Beijing's relic under preservation 1957.

It was originally named Yong'anwanshou Pagoda. Later it was called Cishousi Pagoda because it was built in the Temple of Benevolence and Longevity. The local people also knew it as Linglong Pagoda. The temple was first built and named the Temple of Benevolence and Longevity in the 4th year of the reign of Emperor Wanli of the Ming Dynasty (1576 A.D.). Emperor Qianlong issued an imperial edict to renovate the temple in 1757. It was destroyed by fire during the reign of Emperor Guangxu and only Yong'anwanshou Pagoda and two steles in front of it preserved.

Built of bricks the octagonal multi-eaved pagoda with 13 tiers is exactly 50 meters high. The whole pagoda consists of the base, the body and the spire. The base is divided into two parts, the lower tier of which is a platform made of flagstones and the upper of which is a two-tier shu-mi-tso. On it there are 40 little niches carved with more than 200 figures. Above the shu-mi-tso three tiers of lotus petals further elevate the body. The four front walls of the octagonal body have arches and on each of the four adjacent walls is a window with vertical bars. The arches and windows are all decorated with Guardian Kings and Bodhisattvas.

慈寿寺塔砖雕
Brick Carvings of the Pagoda at the Temple of Benevolence and Longevity

慈寿寺塔
The Pagoda at the Temple of
Benevolence and Longevity

# 燃灯塔
# Randeng Pagoda

　　燃灯塔位于北京市通州区通州城的东北、北运河起点的一侧，是北京地区创建年代最早、保存最完整的佛塔之一，同时也是通州地区重要的标志物。1979年被公布为北京市文物保护单位。

　　燃灯塔建于辽代，是通州著名的古刹佑教寺内的重要建筑。元、明诸代曾予以维修。清代康熙年间曾三次修缮或重修。1900年，八国联军侵占通州期间，佛塔遭到严重的破坏。新中国成立后，政府拨款予以修复。1976年，唐山地震波及北京，塔身受损。1985年，再次修缮，此次修缮，除将塔顶重修外，并将塔刹增高5米，还添加项轮、圆光、仰月宝珠等塔刹构件，同时还将2224枚风铃全数补齐，彻底恢复了佛塔原有的历史容貌。

　　燃灯塔是供奉燃灯佛的宗教建筑。传说燃灯佛是佛祖释迦牟尼的老师，他出生时使身边的一切光明如灯，因此称燃灯佛。

　　燃灯塔是一座砖木结构的实心密檐塔，平面呈八角形，佛塔原高48米，塔围44米。塔身平面呈八角形，四正面辟拱门，四斜面雕饰直棂假窗。塔身中空，塔内置神台，其上供燃灯古佛。塔身之上建造十三层密檐式塔檐，檐下均施仿木结构的重翘单昂五踩砖制斗拱。檐椽及角梁为木制，在每根椽子的端部，都悬挂制作精致的风铃1枚。在每层斗拱的拱眼壁之间，置有佛像1尊，全塔总共有风铃2224枚，神像共计415尊。塔刹是两层八角形须弥座各承仰莲，纵贯铁杵，上置铜质项轮等饰件，有8根铁链连接于塔尖各脊。

Located to the one side of jumping-off points of the northeastern and northern canals in Tongzhou City of Tongzhou District, the Randeng Pagoda is one of the oldest and best-preserved pagodas in Beijing area, and also serves as the landmark of Tongzhou District. It was listed as a Beijing's relic under preservation in 1979.

According to the records of county annals, the Randeng Pagoda first built in the Northern Zhou Period, was the important building situated in the noted Youjiao Temple of Tongzhou District. The pagoda was restored and renovated during the Tang, Liao, Yuan, Ming and Qing Dynasties. In 1900, it was badly destroyed by the Eight-Power Allied Force. After the foundation of the People's Republic of China, the government appropriated funds to restore the pagoda.

Built of bricks and wood the octagonal multi-eaved pagoda was exactly 48 meters high with the circumference of 44 meters. The four front walls of the octagonal body have arches and on each of the four adjacent walls is a window with vertical bars. At the end of each rafter hangs an exquisite bell.

燃灯佛舍利塔基座
The Base of the Randeng Dagoba

燃灯塔全貌
A Panorama of the Randeng Pagoda

# 姚广孝墓塔
# The Stupa for Yao Guangxiao

　　姚广孝墓塔位于北京市房山区青龙湖镇常乐寺村，是北京地区保存最为完整的明代名人墓塔之一。1984年被公布为北京市文物保护单位。

　　姚广孝是元末明初著名高僧，同时又是著名政治家、军事家、史学家、诗人。85岁时病逝，永乐皇帝追封其为荣国公，谥恭靖。清代，曾经对塔进行过程度不同的修缮。20世纪80年代，政府曾两次出资对该塔进行修缮。

　　姚广孝墓塔平面为八角形，塔通高33米，为九级密檐式砖塔。塔座为须弥座式，平座双层勾栏，其下面有仿木构斗栱承托，平座上面是三层仰莲莲瓣雕刻簇拥塔身。塔四正面雕刻仿木隐作隔扇门，四斜面饰雕刻棂花假窗。正门的门楣上嵌石额1块，其上楷书"太子少师赠荣国恭靖姚广孝之塔"。塔身之上建造九层叠涩式塔檐，塔檐每层均系缀铁制风铃。塔刹为铁质，刹杆高3米，杆上筑有项轮和火焰宝珠，并有8条铁链固定于檐角的吻兽上。

　　塔前立神道碑一通，碑额篆书"御制荣国公神道碑"，为明永乐十六年(1418年)所立，碑文为永乐皇帝朱棣亲撰。

Situated at Changlesi Village in Qinglong Town of Fangshan District, the Stupa for Yao Guangxiao is one of the best preserved stupas for notable figures of the Ming Dynasty. It was listed as a Beijing's relic under preservation in 1984.

Yao Guangxiao was a famous senior monk of the last years of the Yuan Dynasty and the early years of the Ming Dynasty, also known as the famous politician, militarist, historian and poet.

Built of bricks the octagonal multi-eaved stupa with 9 tiers, which faces south, is exactly 33 meters high. The base is in the form of a shu-mi-tso, the recession of which is carved with flowers. Above the shu-mi-tso a p'ing-tso of lotus petals further elevates the structure. The four front walls of the octagonal body have doors and on each of the four adjacent walls is a window with vertical bars. Above this, the 9 tiers rise in a slightly bowed profile. Bells hanging from each tier tinkle pleasantly in the wind. The top of the stupa is made of iron.

姚广孝墓塔全貌
The Stupa for Yao Guangxiao

# 镇岗塔
## Zhengang Pagoda

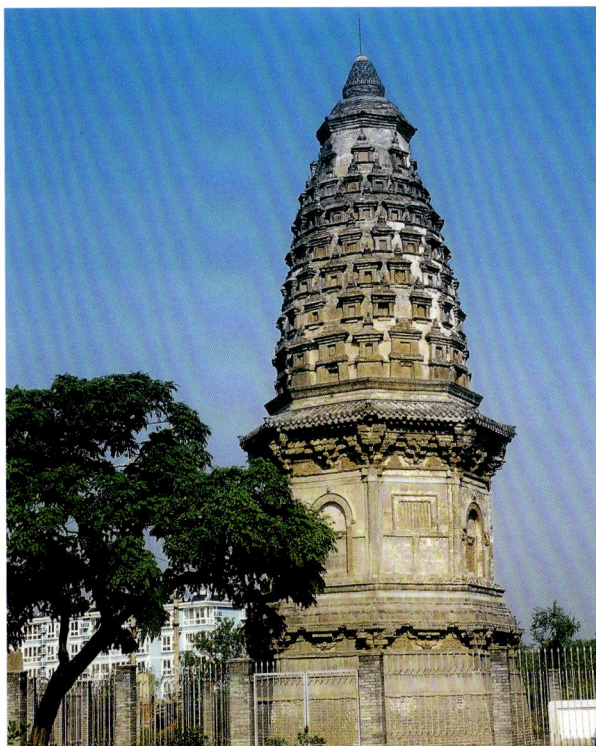

镇岗塔位于北京市丰台区长辛店镇张家坟村，是一座砖结构的实心花塔。1957年被公布为北京市文物保护单位。

镇岗塔始建于金代，明嘉靖四十年(1561年)重修，抗日战争时期，塔底部的一角和塔刹又遭日军炸毁。新中国成立后，先后两次对佛塔进行修葺。

Located at Zhangjiafen Village in Changxindian Town of Fengtai District, the Zhengang Pagoda is a solid flowery pagoda built of bricks. It was listed as a Beijing's relic under preservation in 1957.

First built in the Jin Dynasty, the Zhengang Pagoda was renovated in the 40th year of the reign of Emperor Jiajing of the Ming Dynasty (1561 A.D.). During the Period of Anti-Japanese Aggression, one corner at the bottom of the pagoda and the spire were blown down by Japanese army. The pagoda was renovated twice after the People's Republic of China was founded.

镇岗塔
Zhengang Pagoda

# 良乡塔
## Liangxiang Pagoda

良乡塔又名昊天塔、多宝佛塔，位于北京市房山区良乡城东北的燎石岗上，是北京唯一的一座楼阁式砖塔。1979年被公布为北京市文物保护单位。

良乡塔建造于辽代，光绪二十七年(1901年)被八国联军捣毁，塔前原有接引铁佛1尊，高丈余，亦遭破坏。

Located on Liaoshi Hillock to the northeast of Liangxiang City in Fangshan District, the Liangxiang Pagoda is also known as Haotian Pagoda or Duobao Pagoda and is the only masonry pagoda of pavilion style in Beijing. It was listed as a Beijing's relic under perservation in 1979.

Built in the Liao Dynasty, the Buddha sculptures inside the pagoda were destroyed by the Eight-Power Allied Forces in 1901.

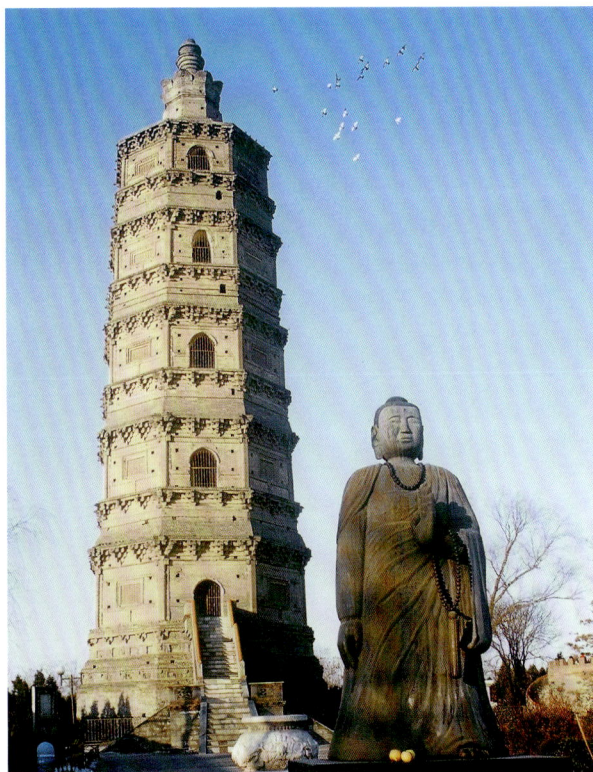

良乡塔
Liangxiang Pagoda

# 白云观
## The White Cloud Daoist Temple

　　白云观位于北京市西城区西便门外西侧，是北京现存规模最大、历史最悠久的道教建筑之一，又是道教全真派三大祖庭之一，有"天下第一丛林"之称号，道观规模宏大，观内文物众多，是道教的一大宝库。2001年被公布为全国重点文物保护单位。

　　白云观前身为唐玄宗敕建的天长观，距今已有1200多年。金代曾两次重修，元代时邱处机曾居于此，改名为长春宫。元末明初，长春宫又遭兵火，明永乐年间敕命重修时，将观址东移，以处顺堂为中心进行扩建，并更名为白云观。清康熙元年至四十五年(1662－1706年)，对白云观进行了大规模维修，奠定今日中路各殿堂的规模。清代光绪中叶，白云观第二十代住持高仁峒与其戒友宫内大太监刘诚印同为慈禧太后所宠信，使全真派的影响达于内廷，盛极一时。建国以后，对白云观进行了几次大规模修缮。

　　白云观建筑坐北朝南，以中路为轴线，配以东、西两路和后花园，共有殿堂17座。

　　中路为主要殿堂所在，最前方为一字青砖影壁，过牌楼、山门，经窝风桥进入中路的四进殿堂，分别是灵官殿、玉皇殿、老律堂、邱祖殿和三清四御阁。其中邱祖殿是白云观建筑群之中心。白云观东路现有殿堂6座和1座三级密檐塔，西路共有殿堂6座。最北端是后花园名云集园，又名小蓬莱，花园由3个庭院连接而成，园内景致清幽，曲廊回环，一别于观前面殿堂建筑规整堂皇的风格，似乎真的像道教描述的洞府仙境一般，别有意境。

牌楼
The Archway

Located at the western side outside Xibianmen in Xicheng District of Beijing, the White Cloud Daoist Temple is one of the largest and most centuries-old Daoist buildings standing here. It is also one of the Three Great Ancestral Courts of the Complete Perfection Sect of Daoism, and is titled *the First Temple under Heaven*. Listed as a national key relic under special preservation in 2001, it is the largest Daoist architectural complex in Beijing and contains numerous cultural relics.

Originally called the Temple of Heavenly Eternity built by the order of Emperor Xuanzong of the Tang Dynasty, the temple has a history of over 1,200 years. In the Jin Dynasty, it was rebuilt twice. In the Yuan Dynasty, Qiu Chuji inhabited here and the temple was expanded as the Eternal Spring Temple by imperial order. During the last years of the Yuan Dynasty and the early years of the Ming Dynasty, the Eternal Spring Temple again suffered from the ravages of war and collapsed once more. When

Zhu Di, Emperor Chengzu, issued an imperial edict to rebuild the temple during the reign of Emperor Yongle of the Ming Dynasty, the temple was moved to the east, enlarged with the Hall of Mildness as its center, and renamed the White Cloud Daoist Temple. In the 1st year of the reign of Emperor Kangxi of the Qing Dynasty (1662 A.D.), large-scale renovations of the temple were conducted. They were completed in 1706. The size of the halls in the middle axis was established.

The main buildings of the White Cloud Daoist Temple, which face south, are divided into the middle, the eastern, and the western axes, and the rear court. There are altogether 17 halls. On the middle axis, from the front to the rear, are the Yi Zi screen wall, the archway, the front gate, the Wuofeng Bridge, the Shrine Hall of Lingguan, the Hall of Jade Emperor, the Laolu Hall, the Hall of Master Qiu, the Hall of Three Pure Gods and Four Guardians and the Yunji Garden.

山门
The Front Gate

窝风桥
The Wofeng Bridge

窝风桥洞
The Wofeng Bridge

灵官殿
The Shrine Hall of Lingguan

玉皇殿
The Hall of Jade Emperor

老律堂
The Laolü Hall

邱祖殿内景
Interior of the Hall of Master Qiu

邱祖殿
The Hall of Master Qiu

三清四御阁
The Hall of Three Pure Gods and Four Guardians

退居楼
The Tuiju Storied Building

戒台
The Ordination Terrace

# 东岳庙
## Dongyue Temple

东岳庙位于北京市朝阳区朝阳门外大街141号，供祀的是东岳大帝众神体系，是道教正一派在华北地区的第一大丛林。庙内大量各具特色的道教塑像和历代碑刻，堪称北京一绝，对研究中国古代道教以及玄教的历史具有重要的参考价值。1996年被公布为全国重点文物保护单位。

东岳庙始建于元代，初名东岳仁圣宫。明代改名为东岳庙。正统十二年(1447年)，在原址基础上全面重建了庙宇，此后的嘉靖、隆庆、万历年间又进行过一些整修和扩建。清康熙三十七年(1698年)，庙遭大火，绝大部分建筑都被烧毁。康熙三十九年(1700年)敕命重建，至康熙四十一年(1702年)修复了东岳庙。乾隆二十六年(1761年)，整座庙进行过重修。道光年间，本庙住持马宜麟四处募化，增筑了东西两座跨院，修建百余间房屋，并创办义学，收容家境贫寒的子弟入学，因而现存建筑均为清代建筑。

东岳庙现占地约71亩，平面分为中、东、西3路，主要建筑都集中在中路的南北中轴线上，从南至北依次有琉璃牌坊、山门(已拆除)、棂星门、瞻岱门、岱宗宝殿、育德殿和后罩楼等共六进院落。东院以居住为主，建筑较为分散，生活气息较浓。西院由供奉各路神祇的小型院落组成，有东岳宝殿(祠堂)、玉皇殿、三皇殿、药王殿、显化殿、马王殿、妙峰山娘娘殿、鲁班殿、三官殿、瘟神殿、阎罗殿以及判官殿等，殿宇的规模都不大，多是由民间人士出资修建而成的。

洞门牌楼
The Archway

瞻岱门匾额
The Board Inscribed with "the Zhandai Gate"

The Dongyue Temple is located at No.141 Chaoyangmenwai Street in Chaoyang District. The Great Dongyue King, the God of Mount Tai, is enshrined in the temple which is the first grand Taoist temple of the Zhengyi Sect in the north China and is titled "the First Temple under Heaven". Numerous statues of the immortals and steles are preserved in the temple. It was listed as a national key relic under special preservation in 1996.

The temple was first built and named Dongyueshengrengong in the Yuan Dynasty. In the 12th year of the reign of Emperor Zhengtong (1447 A.D.), it was rebuilt and renamed Dongyuemiao. Then, it was destroyed and reconstructed several times during the Ming and Qing Dynasties. The current complex dates back to the Qing Dynasty.

The temple covering an area of about 71 Mu is divided into the middle, the eastern, and the western axes. On the middle axis, the main buildings comprise six courtyards one behind another. The principal structures, from the front to the rear, are the glazed archway, the front gate, the Dragon-Tiger Gate, the Daizongbao Hall, the Yude Hall and the posterior shielding storeyed building.

琉璃牌楼
The Glazed Archway

始建于明万历三十年(1602年)，为三间四柱七楼式，歇山顶、黄琉璃筒瓦，脊正中饰火焰宝珠，南北石额各一块，南面书"秩祀岱宗"，北面书"永延帝祚"，相传为明代宰相严嵩所书。牌楼东西原本还各有一座木制牌楼，现已拆除。

岱宗殿
The Daizong Hall

广嗣殿
The Guangsi Hall

赵孟頫墓
Zhao Mengfu's Tomb

碑林
The Forest of Steles

东岳庙的各院落内都立有石碑,最多时达160多块,数量居京城之冠。现存的100多块石碑全为元明清三代的作品,多为历朝修建东岳庙碑记和民间善会石碑,种类齐全,品位高贵,内容丰富,具有较高的艺术和史料价值。其中最著名的是赵孟頫的行书《张天师神道碑》(俗称《道教碑》),风格古朴遒劲,为元代书法艺术的珍品。

三矛殿
The Sanmao Hall

碑亭
The Tablet Pavilion

在福路的两侧，顶覆黄色琉璃瓦，原先放置康熙帝和乾隆帝御笔亲题的石碑。

育德殿
The Yude Hall

为寝宫，面阔五间、庑殿顶、前出抱厦，内饰龙凤天花，与岱宗宝殿相呼应。殿内悬挂清代道士娄近垣所书的"玄妙赞化"匾，原本还供奉东岳大帝和淑明坤德帝后的神像，现改为三官九府像陈列厅。

炳灵公殿
The Binglinggong Hall

后罩楼
The Shielding Building

# 大高玄殿
# Dagaoxuandian

大高玄殿位于北京市西城区景山西街21、23号，是我国唯一一座专供明、清两代皇家御用的道观，建筑极具特色。1996年被公布为全国重点文物保护单位。

大高玄殿始建于明嘉靖二十一年(1542年)，明嘉靖二十六年(1547年)毁于火灾，明万历二十八年(1600年)重修。清雍正八年(1730年)、乾隆十一年(1746年)和嘉庆二十三年(1818年)重新修缮。

大高玄殿坐北朝南，呈南北向，长方形，面积约1.3万平方米。最前方原有牌楼1座(现在的牌楼非原址复建)、习礼亭2座(已拆)，山门后依次排列着琉璃券洞式三孔门，往后为两道过厅式的大高玄门、大高玄殿、九天应元雷坛和乾元阁。大高玄殿是明清最高等级的道观，建筑形式非常多样，独具一格，它使用了最高等级的重檐庑殿顶，别具特色的五花脊，象征"天圆地方"的乾元阁，都是其他道观所不能比拟的，它是北京道观建筑最具特色的一座。

Located at No.21&23 Jingshan Xijie in Xicheng District, Dagaoxuandian used to be a unique imperial Taoist temple for emperors of the Ming and Qing Dynasties. It was listed as a national key relic under special preservation in 1996, the buildings of which are very special.

First built in the 21st year of the reign of Emperor Jiajing of the Ming Dynasty (1542 A.D.), the temple was destroyed by fire in 1547, and was renovated in 1600. Later, it was repaired and renovated several times during the Qing Dynasty.

Facing south, the temple shaped like a rectangle from the north to the south, covers an area of 13,000 square meters. The main buildings, from the south to the north, are the front gate, the glazed arched gate with 3 pairs of doors, the Dagaoxuan Gate, the Dagaoxuan Hall, the altar and the Qianyuan Pavilion. As the highest-grade Taoist temple built during the Ming and Qing Dynasties, the architectural forms are very multiple and special.

牌坊
The Archway

三座门
The Glazed Arched Gate with 3 Pairs of Doors

大高玄殿
The Dagaoxuan Hall

面阔七间，重檐庑殿顶，黄琉
璃筒瓦屋面，檐下都使用斗
拱，绘制最高等级的和玺彩
画，殿前月台围以汉白玉石栏
板，云龙、云鹤柱头,殿前台阶
御路石雕刻云龙图案，雕工简
洁遒劲，月台及大殿台基上均
雕刻螭首，殿内天花藻井非常
大气，且历数百年如同新作。
东西配殿各五间，单檐歇山
顶，绿琉璃瓦屋面。

大高玄殿藻井
The Sunk Panel of
the Dagaoxuan Hall

乾元阁
The Qianyuan Pavilion

位于道观最后方，是一座象征中国古代认为的"天圆地方"宇宙观的两层楼阁，原供玉皇大帝，是清帝祈雨之所。乾元阁上檐圆攒尖顶，覆以蓝琉璃瓦，象征天，悬挂"乾元阁"；下层方形，名"坤贞宇"，覆以

黄琉璃瓦，象征大地，悬挂"坤贞宇"竖匾，阁内部彩画及藻井非常精美，此建筑形式为北京地区孤例。

# 火德真君庙
# The Temple of the Perfect Sovereign of the Virtue of Fire

火德真君庙位于北京市西城区地安门外大街77号，是一座主祀火神的大型道观。1984年被公布为北京市文物保护单位。

火德真君庙的始建年代有两种说法：一说始建于唐贞观六年(632年)；另一说始建于元顺帝至正六年(1346年)。明万历年间，皇宫连续发生火灾，万历皇帝便下令重修火德真君庙，将殿宇改换成琉璃瓦以压火，并增建了一座楼阁。至清乾隆二十四年(1759年)再次重修，将山门及后阁全部更换成黄色琉璃瓦。

火德真君庙的主体现基本保存完整，山门东向，在庙的东南角，门内外原各有一座牌楼，山门外还有旗杆。山门内东西两侧有钟鼓楼各一座(现已不存)，再向西为一灰瓦绿剪边配殿(现已不存)，穿过了这座配殿，庙才转为南北向。前殿为灵官殿，供奉隆恩真君王灵官。第二进是主殿火祖殿，供奉南方火德荧惑星君，即通常所谓之火祖(火神)。此殿顶有一漆金八角蟠龙藻井，精巧无比。火祖殿之北为斗姥阁和万岁景命阁。万岁景命阁两侧的连廊下，各有一掖门，直通庙后一水亭，亭上可观赏什刹海烟波，可惜亭已不存。

Located at No.77 Di'anmenwai Street in Xicheng District, the Temple of the Perfect Sovereign of the Virtue of Fire is a great Daoist temple of worshiping the Fire God. It was listed as a Beijing's relic under preservation in 1984.

There are different sayings about the construction time of the temple. Some people say that it was first built in the 6th year of the Zhenguan reign of the Tang Dynasty (632 A.D.), but some other people think it was constructed in the 6th year of the Zhizheng reign of the Yuan Dynasty (1346 A.D.). Emperor Wanli of the Ming Dynasty issued an imperial edict to rebuild the temple and provided glazed tiles to prevent fires which happened repeatedly in the imperial court. In the 24th year of the reign of Emperor Qianlong of the Qing Dynasty (1759 A.D.), the temple was again rebuilt with yellow tiles added to the roofs of its gates and pavilions.

Today, the main structure of the temple retains its original features. In the south is the Hall of the Perfect Sovereign of Great Kindness, also known as the Heavenly General of the Jade Pivot Fire Office. In the north end lies its main section, namely, the Southern Fire Patriarch Hall in honor of the Perfect Sovereign of the Virtue of Fire, who was also known as the Fire God, or Fire Patriarch.

山门
The Front Gate

藻井
Sunk Panel

# 关岳庙
# Guanyue Temple

关岳庙位于北京市西城区鼓楼西大街149号，是北京的武庙。2006年被公布为全国重点文物保护单位。

关岳庙原是道光帝的第七子光绪皇帝的生父醇贤亲王庙，醇贤亲王庙于光绪十七年(1891年)修建，至光绪二十五年(1899年)建成。民国三年(1914年)北洋政府在后寝祠塑关羽、岳飞像，并祀关岳，称关岳庙。1939年改称武成王庙，简称武庙。大殿改称武成殿，将原关岳殿改称武德堂。堂内北、东、西三面墙壁上镶嵌有配享和从祀武成王的80位历代名将传石刻(现存16块刻石)。

关岳庙坐北朝南，分中、东、西三路，中路为主要殿堂，依次为琉璃砌筑的影壁，山门三间，门外为八字影壁，两旁各有一座琉璃门；中院内东有焚帛炉，西有祭器亭；正殿面阔七间，重檐歇山顶，黄琉璃瓦屋面，檐下施以斗拱、彩画，殿内设神龛，殿前有月台一座，须弥座，白石护栏，龙凤柱头；正殿两侧东西配殿各五间，后寝祠五间。西路有南北两进院落，各有南、北、西房。东路有北房、东房和亭子等建筑。此外庙宇还有神厨、神库、宰牲亭、井亭等建筑，总建筑面积约3000多平方米，是一座融庙宇和祠祀两种特色的庙宇。

Located at No.149 Gulou West Street in Xicheng District, the Guanyue Temple was listed as a national key relic under special preservation in 2006.

The temple formerly served as the ancestral temple of prince Chunxian, the father of Emperor Guangxu. Construction of the temple began in the 17th year of the reign of Emperor Guangxu (1891 A.D.) and it was completed in 1899. In the 3rd year of the Republic of China (1914 A.D.), the temple was used as a place of worshipping Guan Yu and Yue Fei, commonly known as Guanyue Temple. It was renamed Wuchengwang Temple in 1939.

The temple, which faces south, covers an area of 3,000 square meters and is divided into the middle, the eastern, and the western axes. Along the middle axis, the principal structures, from the front to the rear, are the screen wall, the front gate, the main hall and the Houqin hall. The main hall is 7 bays wide with a double-eaved gable-and-hip roof. The roof is covered with yellow glazed tiles with a green edge. On the eastern and western sides of the main hall, there are halls with 5 bays wide respectively.

影壁
The Screen Wall

琉璃门
The Glazed Gate

大殿
The Main Hall

# 牛街礼拜寺
# Niujie (Ox Street) Mosque

牛街礼拜寺位于北京市宣武区牛街88号，是北京规模最大、历史最悠久的伊斯兰教清真寺。1988年被公布为全国重点文物保护单位。

该寺始建于辽代统和十四年(996年)，为辽代入仕的阿拉伯学者那速鲁定所创建。明正统七年(1442年)重修，明朝成化十年(1474 年)奉敕赐名礼拜寺，清康熙三十五年(1696年)时，又进行大规模重修，皇帝并题额"礼拜寺"，所以它是北京唯一一座称为礼拜寺的伊斯兰教清真寺。历经元、明、清各代扩建与重修，使其整体布局集中、严谨、对称。

牛街礼拜寺坐东朝西，意为朝向麦加圣地的方向,是一座典型的中国传统建筑与伊斯兰风格的装饰相结合的建筑群。建筑布局上采用中轴线对称布局，主要建筑有望月楼、礼拜殿、邦克楼和碑亭等建筑。此外，礼拜殿南侧有浴室；邦克楼两侧建有两座方形对称的碑亭；碑亭两侧为南北讲堂各五间，通过游廊与礼拜殿相连；邦克楼东有后殿七间。值得一提的是寺里藏有很多见证伊斯兰教在北京传播、发展的珍贵文物，如白匾、元代的两块阿拉伯文墓碑，以及清代的大铜锅、铜香炉、铁香炉等。

The Niujie (Ox Street) Mosque is located at No.88 Niujie Street in Xuanwu District and is the biggest and oldest in Beijing. It was listed as a national key relic under special preservation in 1988.

The mosque was originally built in the 14th year of the Tonghe reign of the Liao Dynasty (996 A.D.). It was rebuilt in the 7th year of the reign of Emperor Zhengtong of the Ming Dynasty (1442 A.D.) and expended in the 35th year of the reign of Emperor Kangxi of the Qing Dynasty (1696 A.D.). The buildings in the mosque are symmetrically arranged.

The Niujie Mosque, which faces west, the Holy Land of Mekka, presents an aesthetic blend of architecture, reflecting both ancient Chinese palaces and Arabian mosques. The principle structures are the Moon Watching Tower, the Prayer Hall, tablet pavilions etc. On the south of the Prayer Hall is a bathhouse. Many important cultural relics, such as steles from the Yuan Dynasty and cupreous hollowware and incense burner from the Qing Dynasty, are preserved in the mosque.

牌楼和望月楼
The Archway and the Moon Watching Tower

牛街礼拜寺最前方为一座砖石砌筑的一字大影壁，影壁上雕刻有伊斯兰教题材的"四无图"等装饰。影壁后面为一座四柱三楼式牌楼，悬康熙帝书"达天俊路"匾额（"文革"中被毁，现匾额为后仿制）。牌楼后为登高望月之用的望月楼，因为中国的穆斯林入斋和出斋都是以望见新月为标准。望月楼为二层六角形楼阁，黄琉璃瓦绿剪边攒尖顶、宝顶檐下有斗拱和彩画，悬"牛街礼拜寺"匾额一方，和门前的牌坊、影壁交相辉映，组成了一个别具特色的开端。

礼拜殿
The Prayer Hall

坐落在第二进院落，是全寺的主要建筑，坐西朝东，前殿面阔三间，大式歇山顶，前廊用擎檐柱，檐下有斗拱。其后礼拜大殿面阔五间，由一个歇山和一个庑殿勾连搭，再加周围抱厦组成，连前殿总进深达39米，可以容纳上千人同时做礼拜。

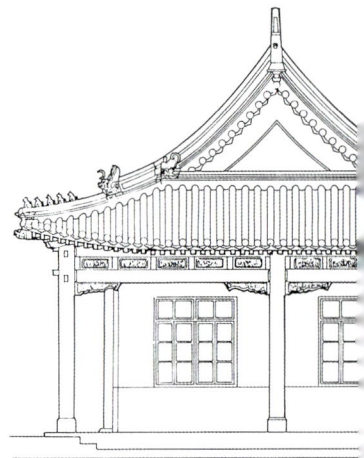

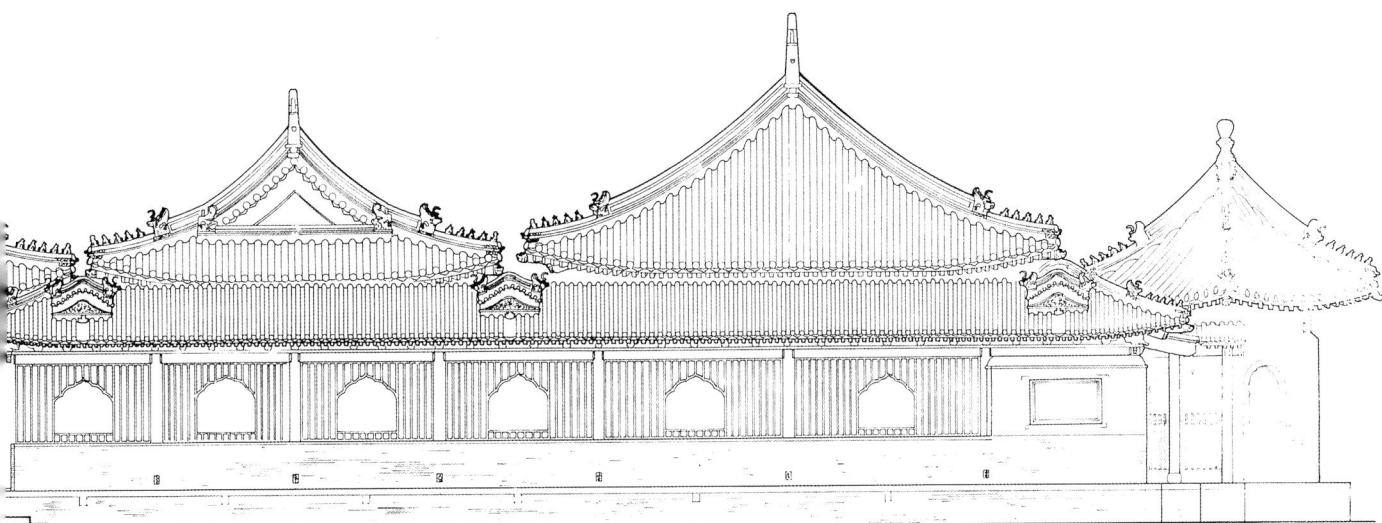

礼拜殿侧立面图
The Side Elevation of the Prayer Hall

大殿入口
The Entrance of the Prayer Hall

礼拜殿内景
Interior of the Prayer Hall

礼拜殿中无任何神像，因为伊斯兰教认为安拉之神无所不在，不赞成供奉塑像，只是在大殿梁柱、花罩和天花板等处饰以博古、花卉和阿拉伯文字等装饰图案，使大殿显得古朴素雅。在大殿的西北侧建有一座木制宣讲台，是聚礼日或节日教长讲经说道之处。殿内的西尽端为窑殿，象征着圣地麦加，六角攒尖顶，南北两侧开券窗，窗棂雕刻阿拉伯文分别为"赐予人类幸福的真主"和"赐予人类恩典的真主"。窗周边饰汉白玉雕花券脸石。

大殿彩画
Coloured Paintings of the Prayer Hall

碑亭
The Tablet Pavilion

# 东四清真寺
## Dongsi Mosque

东四清真寺位于北京市东城区东四南大街13号。又名法明寺，是北京清真寺中著名的"四大名寺"之一，1984年被公布为北京市文物保护单位。

东四清真寺始建年代说法有二：一是始建于元至正六年(1346年)，传说宋元期间有筛海尊哇默定的第三子筛海撒那定在北京东城建立清真寺；二是建于明正统十二年(1447年)，由明代后军都督同知陈友捐资创建。

寺坐西朝东，占地面积约4000平方米。寺门三间，硬山调大脊筒瓦屋面，1920年改建成此样式。门内有南北用房数间。二门为一座中西结合式的建筑，面阔五间，前后出廊，前檐砖砌。门内原来有明成化二十二年(1486年)建的邦克楼(宣礼楼)，二层，方形四角攒尖顶，可惜于清光绪年间被毁。再往后为垂花门一座，门内主要建筑有供做礼拜用的礼拜大殿、南北讲堂、水房和图书馆。礼拜殿前半部为木结构，面阔五间，庑殿顶，筒瓦屋面，后半部的窑殿为无梁殿式穹窿顶结构。礼拜殿三座拱券门门楣上都刻有《古兰经》经文，殿前的南端立有一座明万历七年(1579年)的《清真法明百字圣号》碑。院南北两侧各有五间配殿和三间配房，主配殿均保存明代建筑风格。在南配殿的资料室里，存有各种版本的《古兰经》，最为珍贵的是一本元代手抄本《古兰经》，文字精美，保存完好，堪称国宝，还有埃及国王赠送的图书等珍宝。

The Dongsi Mosque is located at No.13 Dongsinan Street in Dongcheng District, also called Famingsi. It was listed as a Beijing's relic under preservation in 1984.

There are different sayings about the construction time of the mosque. Some people say that it was first built in the 6th year of the Zhizheng reign of the Yuan Dynasty (1346 A.D.), but some other people think it was constructed in the 12th year of the reign of Emperor Zhengtong of the Ming Dynasty (1447 A.D.) by a military officer named Chen You.

The mosque, which faces east, occupies 4,000 square meters of land. The front gate was rebuilt in 1920. The style of the second gate is the combination of eastern and western buildings, which is 5 bays wide with corridors in the front and behind. Originally, there was a minaret inside the second gate built in the 22nd year of the reign of Emperor Chenghua of the Ming Dynasty (1486 A.D.). It fell into ruins during the reign of Emperor Guangxu of the Qing Dynasty, to the south of which stands a drooping flowers gate. The existing buildings inside it are a worship hall and a library. The vestibule of the worship hall is wooden structure. At the rear of the hall are three brick chambers with vaulted ceilings without supporting pillars. The lections of the Alcoran are carved on the hall's arches. There are 5 wing halls and 3 wing rooms respectively on the southern and northern parts of the courtyard. These buildings feature the Ming Dynasty architectural style. In the library located at the southern wing hall, you can find various versions of the Alcoran. The most precious one is a hand-written copy from the Yuan Dynasty. Also there are books presented by the emperor of Egypt.

东四清真寺礼拜殿东立面图
The Elevation of the Prayer Hall of the Dongsi Mosque

大殿内景
Interior of the Main Hall

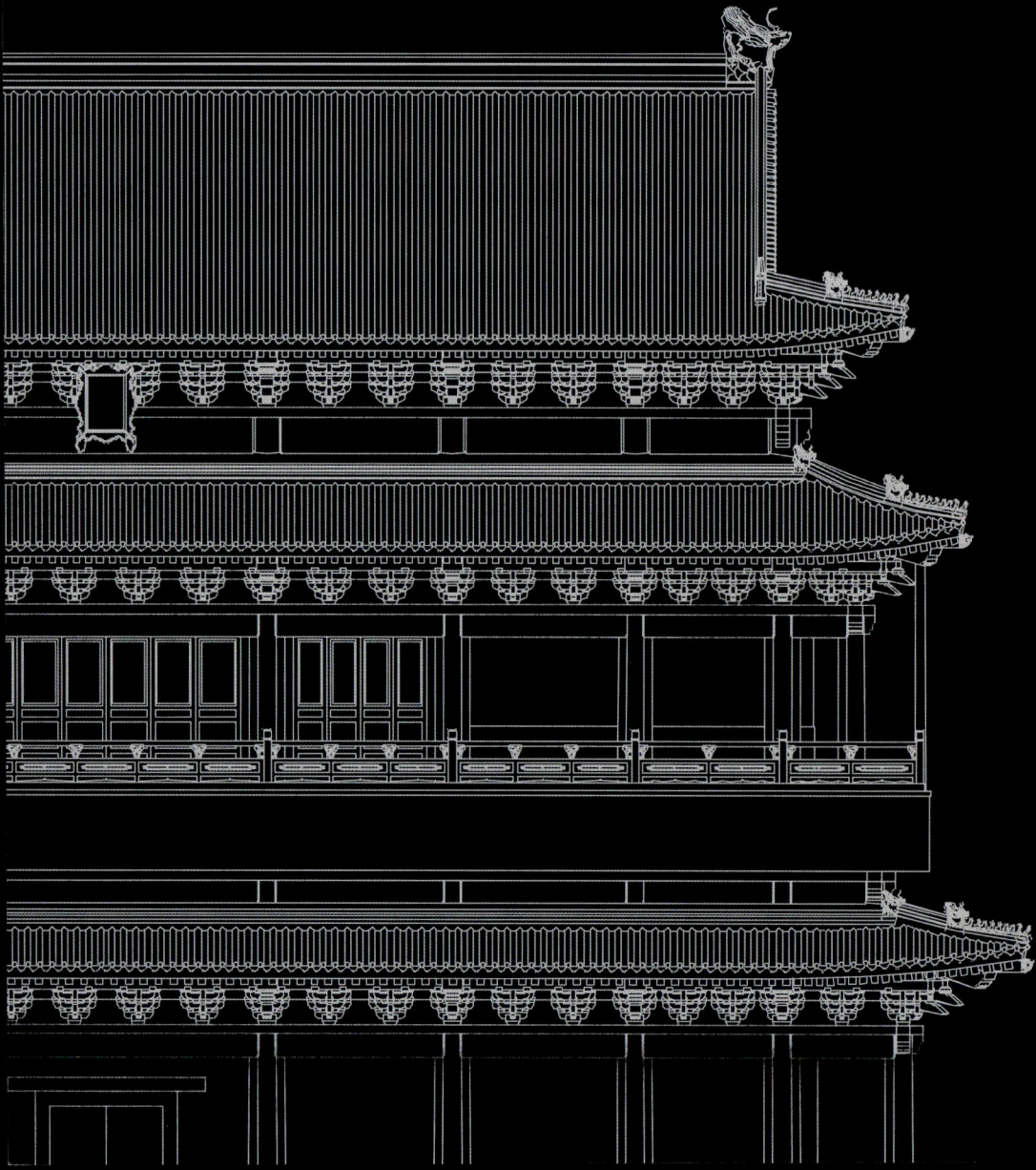

# 北京的城垣

北京自古就是重要的军事重地，这就意味着北京要建立强大的防御体系，并且随着其政治地位的渐趋重要，北京的防御设施逐步完善和宏大。

## 一、北京地区防御设施的发展概况

### 1.商周至隋唐时期

北京作为燕蓟古国，是夏王朝的后裔封地。后来燕的实力强大了就把蓟吞并了，并迁都于蓟城。公元前1045年周灭商后，分封同姓贵族召公奭于北燕。琉璃河遗址的考古发掘，已经发现了此时期的城墙遗址。古城址位于遗址中部，呈长方形，东西长850米，南北长约600米，城墙厚约4米，以土夯筑而成。城墙分主城墙、内附墙和护城坡3部分，城垣外有沟池环绕。地面尚存北城墙和东西城墙的北半部，北墙长829米，东西墙北段尚存约300米，建城年代约在西周初期。虽然这时城墙的主要功用"城以卫君，郭以保民"包含两层含义：一个是防止城内人民逃跑，一是防御作用，但是它却是北京已知最早的防御设施了，还应当是我们已知北京最早的城墙。

到了春秋战国，北京地属燕国，燕国在其境内修筑了南北两条长城，都不经过北京境内，不过也有的学者认为位于今昌平区西北约35公里的西山上的流村乡、老峪沟乡和高口乡的交界处，遗存的南北长约30公里的一段长城是战国末期燕昭王二十九年(公元前283年)修建的。秦统一后废弃。该段遗址大部分残高1.5米、宽2米左右，城台敌楼烽火台等也坍为一堆瓦砾。燕都蓟城成为战国时的名城。

公元前221年，秦在刚刚统一后就派大将蒙恬率领30万大军北驱匈奴。为了便于据守，将旧秦、赵、燕等国长城修缮并连接起来，"因地形，用险制塞，起临洮，至辽东，延袤万余里"[①]形成了著名的万里长城。至汉代，终汉一代北京地区都是北方匈奴的前沿阵地。北齐天保六年(555年)征发180万民夫大规模修筑长城，自幽州夏口(今昌平县居庸关南口)，西到今大同市东北，东至渤海之滨。

隋唐时期是根据军政的重要性划分为州郡，北京地区是隋唐时期的幽州，是全国最大的四大"大都督府之一"，

城墙周围20里以上。唐太宗几次征高丽都是以北京作为一个重要后方，并且在北京建立了悯忠寺(今法源寺)祭奠阵亡将士。历史上著名的"安史之乱"，安禄山也是在范阳(即今北京)起兵，统辖十六州兵马，但是当时唐玄宗重用安禄山的初衷是以此地为控制北方的大基地，说明当时北京的军事地位相当重要。

### 2.辽、金时期

辽代是北京历史地位重要的转折点之一，辽代将北京作为其五京之一的南京。这时南京外围的防御体系主要是居庸关、松亭关、榆林关等关隘。由于辽本身发源于北方，南京以北尽是其掌控，而其所要防控的是南边的宋王朝(所以北京地区此时期修筑长城主要是南防)，北京南边的防线也推移到河北、山东等地。至于南京城的防御体系主要是城墙。金与辽情况一样也发迹于北方而向南发展。金中都城墙是在辽南京城的基础上扩建而成，由于将原城址的东、西、南三面均加以扩大，其城郭便趋向于正方形。中都城周长18690米，城东南角在今永定门火车站西南的四路通；东北角在今宣武门内翠花街，西北角在今军事博物馆南的皇亭子，西南角在今丰台区凤凰嘴村西南角。

### 3.元、明、清时期

元代，北京正式成为全国的首都，由于此时全国已经统一，北方没有了其他少数民族的侵扰，所以北京周边除了旧有的关隘以外，没有再大规模修建新的防御设施。但是，北京的城市防御却建立了规模宏大的大都城墙，将整个城市包围起来。元大都外郭城平面呈南北略长的长方形，东西长6700米，南北长7600米，周长约28600米。其南面城墙在今长安街南侧，南墙在靠近庆寿寺双塔处，稍向外弯曲，以便绕开双塔；其东西两面城墙南段在今东西二环路内侧一线，与明清北京城的东西城墙一致。由于城墙为土质，为了加固城墙和防止敌人攻城时挖掘城墙引起城墙崩塌，在夯土中使用了"永定柱"(竖柱)和"缓柱"(横木)，在早期建筑的木骨泥墙和城墙的排义柱中我们已经可

以见到竖柱，横柱的使用迄今为止发现得最早的是十六国时期的城墙，即城墙上下每隔一段距离就使用横木并排摆放，这样就使得城墙每段都能独立承重，即使敌人把下部挖空也不至于整个墙体崩塌。另外，由于是土墙，为了给城墙防雨，曾使用芦苇披在城墙上面。此外与城墙配套的设施在元代也不断完善，元顺帝至正十九年（1359年），为加强防御，又在大都城的11个城门建造了瓮城和吊桥。

明代是我国历史上又一次也是最后一次大规模修筑长城的时期。明洪武元年（1368年）八月，明军攻入元大都。随即把元大都改称为北平府，并设了地方行政机构北平布政使司，这时，北平虽然不再是全国首都，但是它仍然具有极其重要的军事和政治地位。因为蒙元虽然被明军打败，但是败走的蒙古军队仍然有几十万之众，仍然对明王朝造成极大的威胁，明王朝把北京作为防御蒙古残余势力的基地。明朝立国第一年（1368年），朱元璋就派大将徐达修筑了居庸关等处长城。并且在此后的200余年间，为了防范北方的蒙古和后来兴起的女真等少数民族，明代修筑了我们现在看到的举世闻名的万里长城。明代永乐皇帝朱棣夺取政权后，出于军事、政治等目的迁都北京，为了加强北京的防御，万里长城北京段的建筑者更是花尽心血和智慧修筑了庞大、复杂的长城防御体系。尤其到了明代中叶以后，隆庆年间抗倭名将谭纶、戚继光等著名将领接手北京军务之后，全面整修北京段的长城，使之更是具备了强大的防御能力，所以万里长城北京段是万里长城的精华和代表作。同时为了配合防御，明代还在长城周边设置城堡用于屯军之用。

在加强北京外围防御的同时，北京城的防御设施——城墙也建立了起来，并且不断加固和完善。洪武元年（1368年）八月，明朝大将徐达攻陷元大都。由于元顺帝不战而逃，城市未受到破坏，完整地保留了下来。但是由于城池过大，不利于防守，于是徐达决定将北城墙向南移2.5公里，同时用城砖将城墙外侧包砌起来，以提高其防守能力。1399年，朱棣发动靖难之役，经过3年争战，于1402年夺得帝位，1403年改北平为北京。永乐四年（1406年），

开始筹划迁都北京，在营建宫殿、坛庙的同时将都城南墙向南展拓0.8里筑城，以修建皇城。《大明会典》载："永乐中定都北京，建筑京城，周围四十里，为九门。南曰丽正、文明、顺成；东曰齐化、东直；西曰平则、西直；北曰安定、德胜。正统初更名丽正为正阳，文明为崇文，顺成为宣武，齐化为朝阳，平则为阜成，余四门仍旧"[②]。1436~1445年，明英宗又对北京城进行了第二次增建，主要工程包括：将城墙内侧用砖包砌；建九门城楼、瓮城和箭楼；城池四角建角楼；城门外各立牌坊1座；护城河上的木桥全部改为石桥，桥下设水闸，河岸用砖石建造驳岸。整修之后的京城周长45里，基本成方形，西北缺一角，被附会为女娲补天"天缺西北、地陷东南"之意。但据遥感观测，此处原有城墙痕迹，使内城城墙呈完整的方形。但是这里的地形为沼泽和湿地，不利于地基稳固，因此推测原城墙修筑后不久即被废弃，并修筑斜角的新城墙，将此处割出城外。内城城墙内部为夯土筑成，内侧和外侧均包城砖，通高12~15米、北段和南段厚度大于东西段厚度，平均底厚19~20米，顶厚16米，上有女墙。内城有城门9座、角楼4座、水门3处、敌台172座、雉堞垛口11038个。城外有宽30~60米的护城河，形成了极其坚固的城防体系。北京城建成后，曾多次面临蒙古瓦剌部的入侵，成化十二年（1476年）提出在京城外加筑外城的建议。嘉靖二十九年（1550年）开始修筑前三门外的关厢城（3座独立于城门之外的小城），但由于需要拆毁的店铺民房甚多，民情汹惧，工程不久即停止。明中叶为了加强北京的防御，嘉靖三十二年（1553年）又决定利用元大都土城遗址，四面环绕修筑京城外郭城。最初规划的外城长70里，东西17里，南北18里，设城门11座，敌台176座，西直门外和通惠河设置水闸2处，其他低洼地带设置水关8处。由于工期浩大，于是严嵩建议嘉靖帝改为先修筑南面，将正阳门外的大片繁华工商业市区包入，在包砌完南面后，因用兵频繁，再加上1557年紫禁城大火灾后将财力物力集中于宫殿的重建，因此外城一直没有再筑。1564年增筑外城各城门的瓮城。《大明会典》载："嘉靖二十三年筑重城包京城南一

面，转抱东西角楼止。长二十八里，为七门，南曰永定、左安、右安，东曰广渠、东便，西曰广安、西便。城南一面长二千四百五十四丈四尺七寸；东，一千八十五丈一尺；西一千九十三丈二尺。各高二丈，垛口四尺，基厚二丈，顶收一丈四尺。四十二年增修各门瓮城。"③由此形成的北京城布局此后一直延续了近400年。另外，明代末期为了防御李自成农民军，加强北京的保卫，明崇祯年间，在卢沟桥畔修建了一座专门驻兵拱卫北京城的小城——宛平城。

清代与明代正相反，是中国封建社会历史上少数几个没有修建长城的朝代之一。原因同样是因为清代南北一统，没有受到大规模军事进攻的可能，没有修建长城防御进攻的必要。在北京城的防御上，清代延续明代的城防，即明代修建的城墙，清代只是在明代的基础上进行日常的维修，没有重新构筑其他防御设施。但清代裁撤了皇城的设置，将明代皇城内的大量内廷供奉机构改为民居，将内城的大量衙署、府邸、仓库、草厂也改为民居。同时将内城改为八旗居住区：两黄旗居北，镶黄旗驻安定门内，正黄旗驻德胜门内；两白旗居东，镶白旗驻朝阳门内，正白旗驻东直门内；两红旗居西，镶红旗驻阜成门内，正红旗驻西直门内；两蓝旗居南，镶蓝旗驻宣武门内，正蓝旗驻崇文门内。令汉人迁往外城居住。

## 二、北京现存的古代防御设施

北京的防御设施分为两大部分。

其一，北京地区的防御体系，即举世闻名的万里长城在北京的部分(简称万里长城北京段)以及其附属的关镇，它构成了北京的外围防御体系。万里长城北京段横跨北京北部山区平谷县、密云县、怀柔县、昌平县、延庆县和门头沟区等六个区县，呈半环状分布，全长约629公里(主干线长度为539公里，支线长度为90公里)，其中明代以前的长城长度为73公里，而墙体保存完整的累计长度67公里，仅占全长的10.65%，基本完整的56公里，占8.9%；中等的116公里，占18.44%；坏的95公里，占15.11%；最坏的295

公里，占46.9%。长城线上共有城台(敌台、附墙台及战台827座)，其中好台为391座，占47.3%；坏台436座，占52.7%。城台基本上为方形或长方形，仅禾子涧和南石城等地发现5座圆形台，另外，在怀柔县半城子以北，还发现一座长方形城台筑有坡形屋顶，这种形式的城台极为少见。其中最为雄壮的长城应数八达岭段，而最为险要且敌楼形式最为多样的要数司马台段。此外，北京地区有关口71个、营盘8座。营盘是长城线上用作屯兵、储藏和检修武器的场所，位于墙体内侧，多呈方形或长方形，也有的呈不规则多边形，多分布于远离居民区的长城线上。

其二，北京的城市防御，即城墙、城门，它构成了北京城的防御体系。北京现存有金代、元代和明清城墙，金代和元代城墙为土质，明清城墙为砖质。金代城墙在丰台区凤凰嘴村一带还保存有一些。元代城墙只保存了东、西城墙北段和北面城墙一部分，从今邮电学院小西门，沿旧迹向北走，过蓟门烟树碑，再东转，过德外小关一直到北京旅游学院之南，都可以看到元大都土城遗址。城墙全部用夯土筑成，基部宽24米，高16米，顶部宽8米。在拆除北京西城墙时，在明清城顶三合土之下，发现了元大都土城顶部中心安有排水的半圆形瓦管，顺城墙方向断断续续延续长达300余米。这一发现证明元大都土城的防雨排水也曾采用管道泻水的方法。明清城墙在20世纪五六十年代也基本上全部拆除了，留存下来的也只是内城东南角楼及几百米残墙、内城西南角的一小段、一座城门和一座城门的箭楼了。

注释：
① 汉·司马迁《史记·蒙恬列传》。
② 《大明会典》卷之一百八十七，营造五，城垣，2549页，南京江苏广陵古籍刻印社，1989年。
③ 《大明会典》卷之一百八十七，营造五，城垣，2549-2550页。

# City Defense Installations in Beijing

As an important military base in history, Beijing had to be equipped with a powerful defense system, which was gradually improved and expanded as its political status became increasingly important.

## 1. Evolvement of the Defense Installations in Beijing

### (1) From the Shang and Zhou Dynasties to the Sui and Tang Dynasties

Beijing, where ancient States of Yan and Ji lied, was once a manor of the descendant of the Xia Dynasty. When the State of Yan became stronger, it swallowed up the State of Ji, and moved its capital to the Ji City. In 1045 B.C. when the Zhou Dynasty conquered the Shang Dynasty, the Ji City was enfeoffed to Zhaogong Shi, an aristocrat of the imperial family. The remains of the city wall in this period were found during the archaeological excavation of the site of Liulihe.

Then came the Spring and Autumn and Warring States Periods when Beijing belonged to the State of Yan. Two great walls were constructed in the territory of the Yan State, neither of which passed through the Beijing area. However, some scholars believe that the north-to-south 30 km-long section of great wall in the junction of Liucun Village, Laoyu Village, and Gaokou Village on the Western Mountain, about 35 kilometers northwest of the present-day Changping District, was constructed in the 29th year of the Yan Zhaowang's reign in the late Warring States Period (283 B.C.).

In 221 B.C. when the Qin Dynasty just unified China, General Meng Tian was appointed to lead 300,000 soldiers to drive out the Huns in the north. In order to facilitate the defense, the great walls built by the former states of Qin, Zhao, and Yan were reinforced and connected as whole, which was later well known as the 5,000-kilometer-long Great Wall. In the entire Han Dynasty, Beijing was the forward position for fending off the Huns in the north. In

the 6th year of the Tianbao reign of the Northern Qi Dynasty (555 A.D.), 1.8 million peasants were conscripted to repair the Great Wall in a big scale. The Great Wall started from Xiakou, Youzhou (present-day southern entrance in Juyongguan in Changping County) to the northeast of nowadays Datong City, and to the shore of Bohai Sea in the east.

Beijing, known as Youzhou in the Sui and Tang Dynasties, was a place where one of the four largest military governors settled his mansion, which is surrounded by more than 10 kilometers long of city wall. Beijing served as an important foothold for several expeditions that Emperor Tang Taizong started for conquering Korea. The original intention of Emperor Tang Xuanzong who put An Lushan in a very important position was to take Beijing as the stronghold for controlling the northern region. This shows that Beijing had very significant military position at that time.

### (2) In the Liao and Jin Dynasties

The Liao Dynasty marks an important turning point for the historical position of Beijing. The Liao Dynasty took Beijing (then called Nanjing) as one of its five largest cities. The defense system outside Nanjing City consisted mainly of passes such as the Juyongguan, Songtingguan, and Yulinguan. The city wall of Zhongdu City of the Jin Dynasty was formed following the expansion of the former capital city of the southern city. That is, expanding the former city from the eastern, western, and southern sides, to make the entire city wall look like a square.

### (3) In the Yuan, Ming and Qing Dynasties

By the Yuan Dynasty, the whole country was unified and Beijing became the national capital, so no other large-scale border defense infrastructures than the existing passes were built in the absence of invasion from minor nationalities. Yet, large-scale city wall was built as the city defense installations in Beijing, which encircled the whole Beijing City. The city walls outside Yuan Dadu form a rectangle with 6,700 meters from the east to the west and 7,600 meters from the north to the south, and about 28,600 meters in perimeter. Because the city wall was made of earth, the Yongding pillar (vertical post) and Ren pillar (crossbar) were used during the ramming, in order to fortify city wall and to prevent the city wall from collapse caused by excavation by enemies when they attacked the city. Reed was covered on the city wall to protect it from rain. In addition, the supporting infrastructures had been improved continually in the Yuan Dynasty. In the 19th year of the Zhizheng reign of Emperor Yuanshundi (1359 A.D.), urn cities and suspension bridges were built on the 11 city gates of the Dadu City, in order to reinforce defense.

The Ming Dynasty witnessed a large-scale construction of great walls for a second and also the last time. In the 1st year of the Ming Dynasty (1368 A.D.), Emperor Zhu Yuanzhang sent the general Xu Da to build great wall at the Juyongguan pass and other passes. In the following over 200 years, the present-day famous 5000-km-long Great Wall was built in the Ming Dynasty, in order to fend off the Mongolian invaders and the later-emerging Nuzhen minorities. After Emperor Yongle of the Ming Dynasty seized the regime, he moved the capital to Beijing with military and political considerations. In order to strengthen the defense in Beijing, the builders of the Beijing section of the 5000-km-long Great Wall took every efforts and wisdom to build massive and complex great wall defense fortification. Especially after the mid Ming Dynasty, when two Generals took over the military affairs in Beijing, Tan Lun and Qi Jiguang during the Longqing reign who were famous for defense against the invasion of Japanese pirates, built and renovated the Great Wall in the Beijing section to make them have strong defense capability. Therefore, the Beijing section was the essence and representative of 5000-km-long Great Wall. Fortress was also built at the neighboring areas of the Great Wall

for army cantonment in the Ming Dynasty, in order to reinforce defense.

As the peripheral defense in Beijing was strengthened, the city wall as a kind of defense installations was built, and was reinforced and improved continually. In August of the 1st year of the Hongwu Reign (1368 A.D.), the Ming General Xu Da occupied Yuan Dadu. Because Emperor Yuan Shundi escaped without declaring war, the city was immune from being damaged and well preserved. However, due to the city wall and moat were too large to facilitate defense, General Xu Da decided to move the northern city wall 2.5 Kilometers southward, and also covered the outside city wall with bricks, in order to enhance the city's defense capability. Emperor Zhu Di launched the campaign of Jingnan and seized the throne. In the 4th year of the Yongle reign (1406 A.D.), Emperor Zhu Di began to plan to move capital into Beijing. To build the imperial palace, the emperor moved the southern city wall 0.4 kilometers southward, meanwhile, the palace and temples were built. From 1436 to 1445, Emperor Yingzong built and expanded the Beijing City for a second time. In the 32nd year of the reign of Emperor Jiajing in the middle of Ming Dynasty (1553 A.D.), the outer city wall was constructed encircling the Beijing City to reinforce the defense. In the later period of the Ming Dynasty, in order to defense against Li Zicheng's peasant army, and reinforce the Beijing's defense, the Wanping City, a small city for army cantonment to protect Beijing City, was constructed at the bank of the Lugou Bridge during the Ming Chongzhen reign.

Since the Qing Dynasty unified the northern and southern regions, there is no possibility of large-scale military aggression and thus no necessity to build the defense fortification in the Great Wall. In terms of defense of Beijing City, the Qing Dynasty continued to adopt the city defense system formed in the Ming Dynasty. Only daily repair was made based on those facilities built in the

Ming Dynasty, and no new defense infrastructure was constructed. However, the Qing Dynasty trimmed off the facilities in the imperial palace, and transformed a large number of consecration facilities into residential houses, and some government offices, mansions, and warehouses into residential houses. Meanwhile, the inner city was transformed as the residential area of the Eight Banners (military organization in the Qing Dynasty).

## 2. Ancient Defense Installations in Existence in Beijing

Beijing's defense infrastructure fell into two parts:

Firstly, the defense system in Beijing area consists of the Beijing section of the world famous 5,000 km-long Great Wall and its affiliated passes, which served as the peripheral defense system in Beijing. The Beijing section of Great Wall spans six districts and counties in the northern mountain area, i.e. Pinggu County, Miyun County, Huairou County, Yanqing County, and Mentougou District. The section Changping County takes on a semi-circular shape, with the whole length of about 629 kilometers, of which the section built before the Ming Dynasty was 73 kilometers long, and the most magnificent section should be Badaling section, while the Simatai section was the most strategic place with the most diversified forms of watch towers. In addition, the whole area has 71 passes and 8 barracks. The barracks served as a place along the great wall for army cantonment, military material storage, weapon overhaul, was located in the inner side of the city wall. The barracks, mainly in square or oblong and occasionally in abnormal polygon, were usually distributed along the Great Wall, far away from residential areas.

Secondly, Beijing's defense installations, i.e. city wall and city gate, constituted the defense system of the city. The earthen city walls built in Jin and Yuan Dynasties, and the brick-made city walls built in the Ming and Qing Dynasties remained in existence in Beijing. Some city walls built in the Jin Dynasties remained in the Fenghuanzui Village, Fengtai District. For the city walls constructed in the Yuan Dynasty, only the northern sections of the eastern and western city walls and some part of northern city wall remained. By and large, the city wall of the Ming and Qing Dynasties were dismantled in the 1950's and 1960's, and only the turrets and several hundred meters of remnant walls in the southeastern inner city, and the small section and a city gate, and a embrasure watchtower in the southwestern inner city are still in existence.

# 万里长城——八达岭
# The Great Wall in Badaling Section

　　万里长城——八达岭位于北京市西北60公里的延庆县境内燕山沉降带西端，建筑在海拔600米～1000米的山脊之上。八达岭长城是我国古代伟大的防御工程万里长城的一部分，是明长城中保存最好的一段，也是最具代表性的一段。1961年被公布为全国重点文物保护单位，1987年列入世界文化遗产名录。

　　八达岭地处关沟北端，元代时又名北口。其地势居高临下，形势险要，而且又是居庸关的门户，是防守的主要阵地，因此成为历代兵家必争之地，早在战国时期此地已构筑了长城防御工事。汉代在这里设置了军督和居庸两座关城。北魏修造的"畿上塞围"长城，西起黄河，东至上谷军督山(即今八达岭)。

　　八达岭是一座关城，建于明弘治十八年(1505年)，砖石砌筑，城高7.5米、厚4米，呈东窄西宽的梯形，面积约5000平方米。关城有东、西两座关门，东门门额书"居庸外镇"，西门门额书"北门锁钥"，两门均为砖石结构，条石基础，砖券洞，券洞上为平台，台上四周砌垛墙，并于南北垛墙上各开一豁口，有登城马道可以下到城中。西门南北两侧连接着高低不一、曲折连绵的长城，该段长城建造坚

固，充分借览了历代修长城"因险制塞"的经验。平均高约7.8米，墙基宽约6.5米，顶宽5.8米，全部砖石结构。城墙顶部由三四层砖铺砌而成，在山势陡峭的地方，城墙顶上修筑梯道，八达岭长城上的梯道长达千米。城墙内侧设置宇墙，外侧设置垛口，垛口中间有方孔为瞭望孔，下有一洞为射洞，用以瞭望和射击敌人。墙面上有排水沟和吐水嘴。为了加强防御，在城墙上每隔三五十米或一二百米就建有一个墙台或敌台。墙台是跨墙而筑的平台，多为单层，较宽大，四周也有垛口，是巡逻放哨的地方；敌台是一种临战的工事，结构复杂，形势多样，多为上下两层，上层有垛口可以瞭望射击，下层有券洞，可供住宿，是守望和住宿的地方。此外，从八达岭长城西望，可以看到许多土石筑起的烽火台，是古代传递军情的工具。

　　八达岭关城东门外路旁有一块巨石，据说过去天气晴朗的时候站在上面，可以看到北京城，因而被称作"望京石"。在关城入口处有几门明代的大炮，其中最大的一门名为"神威大将军"，1638年铸造，长2.85米、周长1.05米。其余还有四门牛腿小炮和235枚炮弹，都是1957年修整长城时出土的。

关城西门石额
The Board over the Western Gate of the Fort

八达岭长城
The Great Wall in Badaling
Section

Located to the west of Yanshan Mountains in Yanqing County 60 kilometers to the northwest of Beijing, the Great Wall in Badaling Section was built on the ridges being 600-1,000 meters above sea level. Badaling, being a part of the Great Wall, an ancient great Chinese military defense project, is the best-preserved and representative section of the Great Wall of the Ming Dynasty. Listed as a national key relic under special preservation in 1961, the Great Wall in Badaling Section was also inscribed in the World Heritage List of the United Nations Educational, Scientific and Cultural Organization (UNESCO) in 1987.

Badaling is located to the northern end of Guangou, a narrow valley lined by precipitous mountains on both sides, made the spot a defile, also called Beikou in the Yuan Dynasty. Because of the importance of Badaling, it had been the place that the military fought against for. The defense project of the Great Wall was built here in the Warring States Period. During the Han Dynasty, two fortifications were built in this area, namely Jundu and Juyong. In the Northern Wei Dynasty, "the Great Wall around Ji (the capital)" was built. The wall started from the Yellow River in the west to Jundu Mountain in the east.

Built in the 18th year of the reign of Emperor Hongzhi of the Ming Dynasty (1505 A.D.), the Badaling Fort, which was built with stone slabs, is 7.5 meters high and 4 meters

八达岭长城北路
The North Route of Badaling Section

thick, covering an area of 5,000 square meters. The fort has a gate in both the east and west. There is a board over the eastern gate with four characters for "the Outer Post of Juyong". On the board over the western gate are characters for "the Strategic Gateway of the North". Two gates are made of stones and bricks, on which is a platform surrounded by buttresses. There are two breaks on both sides of the northern and southern buttresses connecting bridle paths. The Great Wall runs up along the southern and northern ridges of the mountain. This section is 7,090 meters long. The wall is 7.8 meters high on the average. The base of the wall is about 6.5 meters wide on the average, while the top of the wall is about 5.8 meters wide on the average. Outside of the wall are buttresses. A hole between every two buttresses is called watch-hole, and peepholes under the holes called embrasures. Inside of the wall, there are low walls with one meter high called parapets, which can be used as railings. There is a scroll door not far from the inside wall, with is a stone ladder for climbing up and down. For the purpose of defense, a watching tower was built every 30-50 meters or 100-200 meters.

八达岭长城南路，南七楼至南十楼
The South Route of Badaling Section, from No.7 Tower to No.10 Tower in the South

八达岭长城北路，北四楼
The North Route of Badaling Section, No.4 Tower in the North

八达岭长城北路，北六
楼至北七楼
The North Route of
Badaling Section, from
No.6 Tower to No.7
Tower in the North

城墙垛口
The Crenel on the Wall

八达岭长城烽火台
The Beacon Tower in Badaling Section

# 长城——司马台段
# The Great Wall in Simatai Section

  司马台长城位于北京市密云县东北部的古北口镇境内，距北京120公里。司马台长城是古北口长城的一部分，是明代著名将领戚继光、谭纶等精心设计、重点增修的城段，也是我国唯一一段保留明代原貌的古长城。这段长城已被联合国教科文组织确定为"原始长城"。2001年被公布为全国重点文物保护单位。

  司马台长城以司马台水库为中心，向东延伸城墙和16个敌台，城墙延伸总长2400米，向西延伸城墙和18个敌台，城墙延伸3000米，城墙总长5400米，计有35座敌台(有一个沉于司马台水库中)，以险峻、敌台城楼形式多样、保存完整而著名。司马台长城修筑年代可分为：单边墙修筑于明成化年间，天梯形边墙修筑于明嘉靖二十四年(1545年)，西七台、东二台和东四台创建于明隆庆四年(1570年)和隆庆五年(1571年)。其余大部分城墙和敌台修建于明万历年间。

  司马台长城属于明长城中最坚固的蓟镇长城，由于地形复杂和修筑者的智慧创造，使这段长城风貌独特。司马台是古北口东麓的重要隘口，称为司马台暖泉口。口门两侧山势险峻，起伏变化比较大。这里的长城及防御工事，对捍卫古北口、防止外族侵扰起着举足轻重的作用。此段长城沿山脊建造，最险处城墙地处悬崖的山尖上，仅宽40厘米。其东部望京楼，是海拔986米古北口长城东部的制高点。在如此短的距离内，司马台长城拥有如此形式多样、变化多端的城墙和敌台是整个万里长城中少见的一段，是中华民族智慧和创造力的代表作品。

司马台长城
The Great Wall in Simatai Section

Located in Gubeikou Township to the northeast of Miyun County 120 kilometers from Beijing, the Great Wall in Simatai Section is a part of the Great Wall in Gubeikou Section. Carefully designed and built by Qi Jiguang and Tan lun, famous generals of the Ming Dynasty, this section still keeps the original appearance of the Great Wall of the Ming Dynasty. Confirmed by UNESCO as "the Original Great Wall", it was listed a national key relic under special preservation in 2001.

In the central position lies the Simatai Reservoir flanked by walls and watchtowers on the eastern and western sides. The wall on the east is 2,400 meters long with 16 watchtowers, while the wall on the west is 3,000 meters long with 18 watchtowers. The wall is totally 5,400 meters long with 35 watchtowers. The section of Sky Bridge was built during the reign of Emperor Chenghua of the Ming Dynasty. The section of Sky Ladder was built in the 24th year of the reign of Emperor Jiajing (1545 A.D.). Besides, most walls and watchtowers were built during the reign of Emperor Wanli of the Ming Dynasty.

This section was ingeniously created and uniquely designed with novel structure and diversification. It concentrates many characteristics of the Great Wall in one section. Simatai is an important pass of the eastern foot of Guheikou. This section was built along mountain ridges. The narrowest is no more than 40 centimeters wide, on both sides of which are sheer precipices. The Overlooking Beijing Tower, which is 986 meters above the sea level, is the highest place of Simatai.

司马台长城三眼敌楼
The Three-Balistraria Tower

司马台长城三眼敌楼线图
The Plan of the Three-
Balistraria Tower

司马台长城敌楼内部
Interior of the Watchtower

司马台长城
The Great Wall in Simatai Section

通向望京楼的"天梯"
The Sky Ladder Style Wall toward theTower for Looking at the Capital

司马台长城障墙及敌楼
The Barrier Wall and the Watchtower

司马台长城望京楼
The Tower for Looking at the Capital

# 居庸关 云台
# Juyongguan and Cloud Terrace

　　居庸关在北京城西北100余里，是万里长城的一个重要关口，为古代北京西北的重要屏障。

　　秦朝即有居庸关之名，后历代均在此修筑长城。现存的居庸关关城和长城，是明朝初期建筑的。明太祖灭元，为了防御元顺帝卷土重来，派大将军徐达修筑居庸关等关口，垒石为城。正统末年英宗朱祁镇亲征瓦剌，兵败被俘。兵部尚书于谦等拥立代宗朱祁钰继帝位，建元景泰，并将居庸关关城扩大加固。现在的居庸关门额上，还有"景泰五年（1454年）立"的题款。

　　居庸关两旁高山屹立，中间是一条长达40余里的关沟，关城建在沟的中间，东西两面依山筑城，形势险要，为历代兵家必争之地。又因山峦间山花野草，葱茏郁茂，登高远眺，好似碧波翠浪，景色异常幽美，因此，很久以前就被称作"居庸叠翠"而列为燕京八景之一了。

　　居庸关还有南北两个外围关口。南面的叫南口，是关沟的入口，北口即是八达岭，历代都将此作为防守的重镇。元时虽不存在北部侵扰之患，也在这里设万户府，统领各部卫军3000人及南北口各地卫军，以检查过往客商和缉拿盗贼。明代在这里设"卫"，经常整修关城，整饬军备，严加防守。

　　居庸关关城的中心，有一座白色的大理石台子，叫做"云台"，为居庸关仅存的文物建筑。1961年，居庸关云台被公布为全国重点文物保护单位。

居庸关全貌
A Panorama of Juyongguan

Located about 50 kilometers to the northwest of Beijing, Juyongguan was an important pass of the Great Wall and a natural barrier to the northwest of ancient Beijing. Two characters of Juyong dated back to the Qin Dynasty. Then, each dynasty built the Great Wall here. The Juyongguan and the Great Wall in this section extant was built in the early Ming Dynasty. In the first year of the Hongwu reign, General Xu Da conquered Dadu. Later, some remaining forces launched frequent southward harasses. In order to guard against invasions, Emperor Hongwu ordered Xu Da to build a pass at Juyongguan,

over the southern gate of which, a board with three characters for Juyongguan written on it, was built in the 5th year of the reign of Emperor Jingtai(1454 A.D.).

As early as the Jin Dynasty, "Overlapping Verdant Scenes of Juyong", known for its dense trees and blooming flowers, was one of the eight scenic spots of Yanjing. Looking into the distance from the peak, one can enjoy the magnificent scenery.

Juyongguan as a pass ranges from Nankou in the south to Badaling in the north. In the center of Juyongguan lies a white terrace made of marbles, called Cloud Terrace.

云台及石雕线图
The Plan of the Cloud Terrace and Stone Carvings

云台
The Cloud Terrace

云台原是一座过街塔的台座，元朝末代皇帝惠宗命人建造，建于元至正二年至五年(1342－1345年)。当时，台上建有3座并排的白色藏式佛塔。过街塔北面还有1座大庙，名"永明寺"。过街塔是"永明寺"的一部分。但台上的佛塔在元末明初已经毁坏。明正统年间曾于台上建寺名"泰安"，寺清初尚存，顾炎武《昌平山水记》有载，后也被毁。今台面上尚遗留有柱础。

云台是座用大理石砌成的长方形台子，上顶25.21米、进深12.9米、下基宽26.84米，进深17.57米，台高9.5。台基正中有一个门洞贯通南北。洞长17.57米，可以通行车马。门洞上部为半个八角形，形式颇为别致。台顶四周围有白石护栏，栏柱下面各有一个排水龙头，栏杆地栿之下的平盘下，雕刻璎珞串珠、兽面等纹饰。

居庸关云台有着很高的历史和艺术价值，其雕刻手法圆润流畅。造型别致，图案精美，是一个巨型石雕艺术品，是现存元代雕刻艺术和建筑技术的优秀代表作。

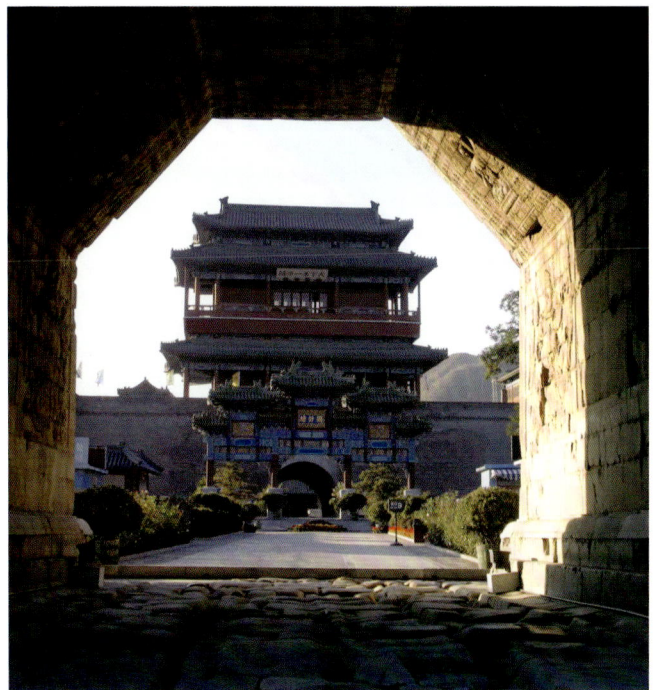

从云台望关城
Looking at the Gate Tower of Juyongguan from the Cloud Terrace

居庸关云台券洞侧壁雕刻四大天王像之一
One of Four Hearenly Kings Carved on the Wall of the Vaulted Gateway of the Juyongguan Cloud Terrace

天王像(局部)
Section of the Carving of Heavenly King

居庸关云台拱券顶雕刻——十方佛
Carvings on the Vault of the Gateway of
the Juyongguan Cloud Terrace: Buddhas
for Ten Directions

居庸关云台拱券洞顶部雕刻
Carvings on the Vault of the Gateway of the Cloud Terrace

云台拱券洞顶部正中平面上，刻着五个"曼荼罗"，两侧斜面上刻着
十尊坐佛，十佛之间遍刻小佛。

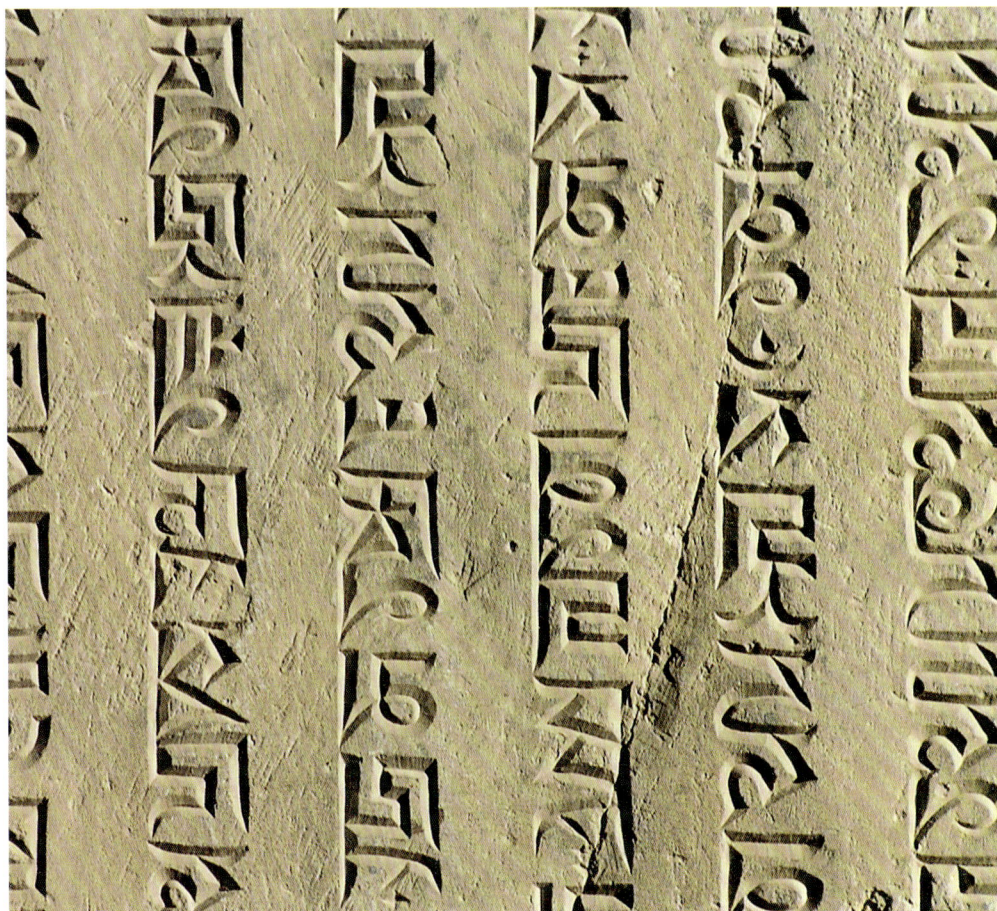

居庸关云台八思巴文石刻
Characters Inscribed in the Basiba Language on the Wall inside the
Gateway of the Cloud Terrace

云台内壁两侧用六种文字刻有同样内容的经咒和造塔功德记，两侧的
排列方法和内容都一样。六种文字为梵文(古尼泊尔文)、八思巴文、
藏文、维吾尔文、西夏文和汉文。这六种文字同时刻在一起，在我国
古代石刻中仅此一例，它也可以作为我国元代各族人民的文化交流和
互相往来的证明。在汉文造塔功德记的末尾有至正五年(1345年)的款
识，说明了过街塔的建成年代。

# 正阳门
# Zhengyangmen

正阳门位于天安门广场南侧，为明清北京城内城的正南门，因其位于紫禁城与皇城的正前方，故俗称前门。正阳门是北京城内唯一保存完整的城楼、箭楼。1988年被公布为全国重点文物保护单位。

正阳门始建于明初永乐十七年（1419年），明成祖营建北京城，将元大都城南城垣和丽正门向南移建0.8公里，仍沿元旧称"丽正门"，永乐十九年（1421年）竣工。尔后，重建城楼、增建箭楼、瓮城及闸楼并于正统四年（1439年）竣工，更名"正阳门"。正阳门因位于皇宫正前方，均较其他八门规制有所不同，是内城九门中最雄伟高大的一座。明清时期，正阳门多次遭受火灾并修复。1900年，八国联军入侵北京，正阳门城楼、箭楼被毁，1906年参照崇文门城楼、宣武门箭楼规制修复完成。1915年为改善正阳门交通堵塞状况，对城门进行了改造。正阳门城楼、箭楼于1949年建国后进行多次修缮，并对外开放。

正阳门城楼坐北朝南，建筑面积5091平方米，为重檐歇山三滴水楼阁式建筑，绿剪边灰筒瓦屋面，坐落在宽95米，厚31.45米，高14.7米的砖砌城台上，城台上窄下宽，有显著的收分，城台南北上沿各有1.2米高的宇墙。城台门洞为拱券式，开在城台正中，五伏五券式，内券高9.49米，宽7.08米，外券高6.29米，宽6米。楼台整体通高42米，为北京各城门中最高者。城楼有上下两个功能层和中间的一个结构层支撑上层平座，两层均面阔七间，进深三间，带周围廊，上层四角立擎檐柱，檐下均施以斗拱。一层为朱红砖墙，明间及两侧山面各有实榻大门一座，二层前后檐装修为菱花假隔扇门，城台内侧设马道一对。

箭楼位于城楼南端，坐北朝南，建筑面积4664平方米，为砖砌堡垒式建筑，城台高约12米，正中设券门与城门相对，且券门在明清时期只供皇帝出入。南面券门五伏五券，北面为1915年新加平台门洞，三伏三券。门洞南侧宽10米，北侧宽12.4米，门洞内设"千斤闸"。箭楼面阔七间，北侧有三层抱厦，面阔五间，重檐歇山屋面，顶覆绿剪边灰筒瓦，檐下均施以斗拱。箭楼东、西、南三面辟箭窗，南面四层每层13孔，东西各4层每层4孔，抱厦左右各5孔（1915年改造时新增4孔），共有箭窗94孔。整座箭楼通高38米，为北京所有箭楼中最高者。

已拆除的瓮城平面呈长方形，南北长109米，东西宽85米，东北、西北二内角为直角，东南、西南二外角为抹角。瓮城东西两侧开券门，券门内有"千斤闸"，券门上建有闸楼3间，其外侧正面设箭窗两排共12孔，内侧正面辟过木方门。

正阳门箭楼南立面
The Elevation of
the Embrasure
Watchtower of
Zhengyangmen

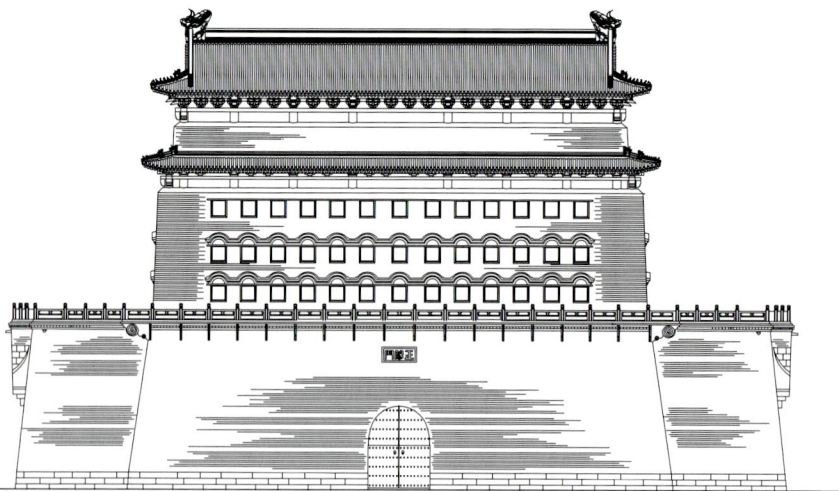

正阳门箭楼
The Embrasure Watchtower of
Zhengyangmen

Located in the central section of the inner city's southern wall, Zhengyangmen was the frontispiece of the inner city in the Ming and Qing Dynasties. It is also known as Qianmen because it stands in the front of the Imperial City and the Forbidden City. As the only best-preserved, the gate tower of Zhengyangmen and the embrasure watchtower were listed by the State Council as a national key relic under special preservation in 1988.

Zhengyangmen was first built in the early Ming Dynasty. The city wall of the Ming Dynasty was based on the earth wall of the Yuan Dynasty. The city wall in the south and Lizhengmen was expanded 0.8 kilometers southward. The construction was completed in the 19th year of the reign of Emperor Yongle (1421 A.D.). Thereafter, the embrasure watchtower and the urn city were built and completed in the 4th year of the reign of Emperor Zhengtong (1439 A.D.). The gate tower of Zhengyangmen was rebuilt and renamed Zhengyangmen. Located in front of the Imperial Palace, Zhengyangmen was the highest and the most magnificent among the nine gates of the inner city. In 1900, the gate tower was destroyed and the top of the embrasure watchtower was burn down by the Eight-Power Allied Forces. They were not renovated until 1906.

The 42-meter-high Zhengyangmen consists of the gate body and the gate tower. The gate tower has a double-eaved gable-and-hip roof covered with grey tube-shaped tiles with a green edge. It stands on the 95-meter-wide, 31.45-meter-thick and 14.7-meter-high gate body made of bricks. The arch gateway cuts through the middle of the gate body, and on both sides are bridle paths.

Located to the south of the gate tower, the 38-meter-high embrasure watchtower made of bricks is in the shape of a fort. The arch gateway cuts through the middle of the 12-meter-high gate body, and was used only for emperors' passage in the Ming and Qing Dynasties. The 62-meter-wide and 20-meter-deep gate tower is 7 bays wide with 5 bays on its north. The embrasure watchtower with 94 arrow windows has a double-eaved gable-and-hip roof covered with grey tube-shaped tiles with a green edge.

正阳门城楼
The Gate Tower of Zhengyangmen

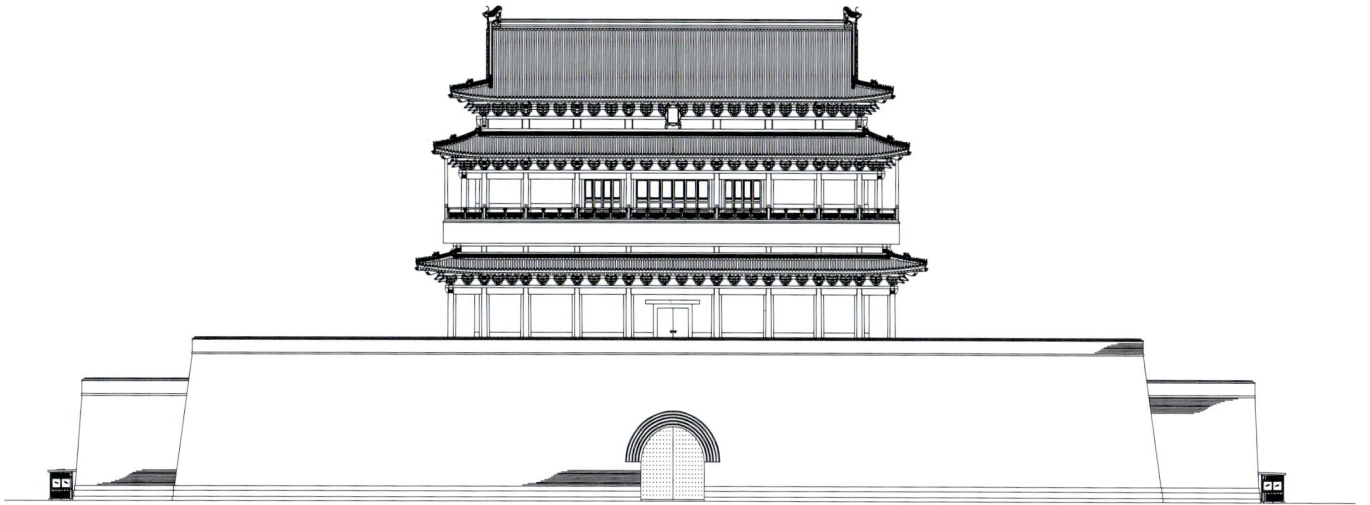

正阳门城楼南立面
The Elevation of the Gate Tower of Zhengyangmen

山花
Coloured Paintings

天花
The ceiling

# 北京城东南角楼
# The Southeastern Corner Tower

北京城东南角楼位于内城东、南垣交角处，今建国门南大街和崇文门东大街相交处的内侧，是北京唯一保留下来的角楼，也是全国现存最大、最早的城垣角楼。1982年被公布为全国重点文物保护单位。

北京城东南角楼始建于明正统元年（1436年），正统四年（1439年）竣工。明嘉靖、隆庆及清乾隆朝均有不同程度的修缮。1900年，八国联军中的美、俄军的炮火几乎将东南角楼彻底摧毁，清末曾进行修补。

角楼建于突出城墙的方形台座上，台高12米，底边长39.45米，上边长15米，楼高17米，通高29米，整座楼的建筑面积为701.3平方米。角楼沿城台外缘转角建起，四面砖垣，采用呈两翼对称并带后抱厦的曲尺形平面，具有显著的对外防御的设计意图。角楼为重檐歇山式建筑，顶覆绿剪边灰筒瓦，两条正脊于转角处相交成十字，檐下均施以斗栱。楼体外侧东南两长面各辟箭窗四层，每层14孔，西、北两窄面各辟箭窗四层，每层4孔，共144孔。楼体内侧随主楼各出抱厦，亦相连成转角房，辟二门，门上方各设直棂窗三扇。楼内立有20根金柱支撑梁架，是角楼的主要承重构件，东南侧铺设三层楼板，局部形成4层空间，并于西北两侧留有各层贯通的大空间。角楼城台内侧筑马道一对。北京城东南角楼整座建筑形式特殊，是我们今天尚能见到的明清城防设施的特征。

Located at the southeastern corner of the inner city, namely the intersection between Jianguomennan Street and Chongwenmendong Street, the Southeastern Corner Tower is the only preserved corner tower of the city in Beijing, and is the earliest and the biggest corner tower of the city wall existing in China. It was listed by the State Council as a national key relic under special preservation in 1982.

First built in the 1st year of the reign of Emperor Zhengtong of the Ming Dynasty (1436 A.D.), the corner tower was completed in April, 1439. It was repaired and renovated several times during the Ming and Qing Dynasties.

Covering an area of 701.3 square meters, the corner tower is located on a rectangular platform jutting out from the exterior of the city wall with 29 meters in height. The platform is 12 meters high with a 39.45-meter-long bottom and a 15-meter-long top. The tower is built from the corner of the exterior city platform and has a curved-ruler shape. It is filled with tiled walls on all four sides. The double-eaved gable-and-hip roof is covered with grey tube-shaped tiles with a green edge. Its two ridges intersect at the corner to form a cross. The corner tower has 144 arrow windows for shooting arrows.

北京城东南角楼
The Southeastern Corner Tower

# 德胜门箭楼
## Deshengmen Embrasure Watchtower

德胜门箭楼位于内城北垣西侧，即今德胜门立交桥北侧，是明清北京城的重要城防工程之一，也是重要历史建筑遗存之一。2006年被公布为全国重点文物保护单位。

德胜门建于明正统元年至四年（1436－1439年），为北京内城的九座城门之一，是由城楼、瓮城和箭楼等组成的群体城防建筑，箭楼是护卫城门的军事堡垒，在城市的防御上起着重要作用。

德胜门箭楼矗立于瓮城之上，坐南朝北，重檐歇山顶，绿琉璃瓦剪边灰筒瓦屋面，前楼后厦合为一体，平面呈倒"凸"字形。北为正楼，面阔七间。南接庑座五间，前檐开座过木方门，四檩单坡顶，檐下施以斗拱，上檐枋额、角梁、斗拱都绘以旋子彩画，是典型的清官式建筑。箭楼上下四层，每层都辟有箭窗，共计82个，作为守城时对外射击的孔道。

Locate at the western side of the northern wall of the inner city, namely the northern side of Deshengmen Crossroads, the Deshengmen Embrasure Watchtower was one of the very important city defense projects in the Ming and Qing Dynasties. It was listed as a national key relic under special preservation in 2006.

As one of the nine gate towers in the inner city, Deshengmen, which consists of the gate tower, the urn city and the embrasure watchtower, was built from the 1st year to the 4th year of the reign of Emperor Zhengtong of the Ming Dynasty (1436-1439 A.D.). Standing in front of the gate tower, the embrasure watchtower, a military fort for guarding the city gate, used to play an important role in city defense. Now, only the embrasure watchtower and the wall of the urn city in section remain.

The embrasure watchtower with 82 arrow windows, which faces north, has a double-eaved gable-and-hip roof covered with grey tube-shaped tiles with a green edge.

德胜门箭楼远景
The Deshengmen Embrasure Watchtower Viewed from Far

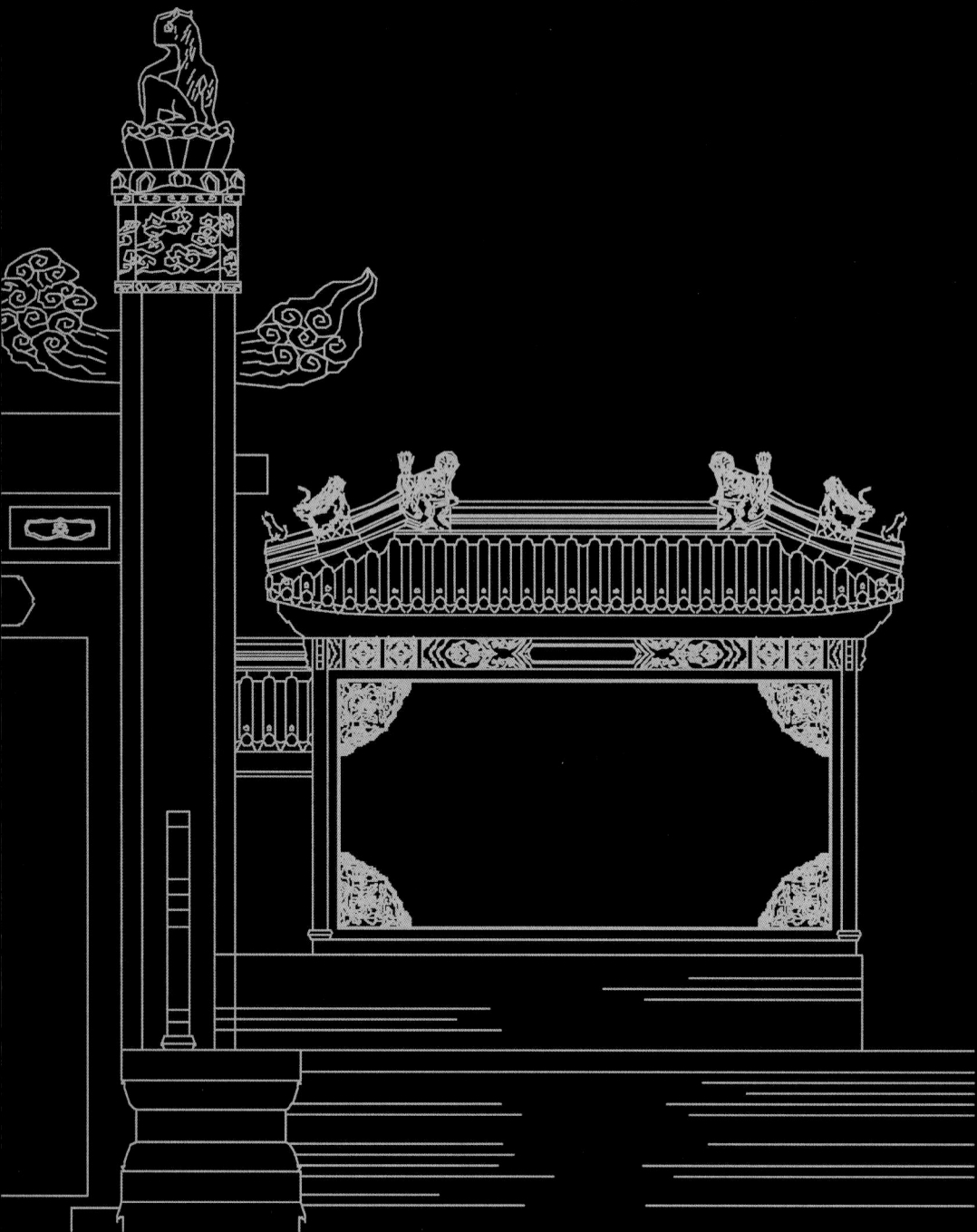

# 北京的陵墓

北京是一座有着3000年历史的文化古都，由于所在的地理位置和环境的缘故，远在春秋战国时期，统治阶级就将这里视为风水宝地。数千年来，这里聚集了一代又一代的王侯将相，并选择这里作为死后的安息之所。特别是封建社会后期金、元、明、清四代，北京成为全国的政治、经济和文化的中心，由于政治地位的提升，北京的墓葬级别也由最高的王爷墓提高为帝王陵墓，并出现了由帝王陵组成的特大型陵墓群，北京房山区的金代陵寝和昌平区的明代十三陵，都是北京陵墓的最高等级的代表作。

## 一、北京最早的王侯墓——琉璃河燕国王侯墓葬

北京房山区琉璃河商周遗址，位于北京房山区琉璃河董家林村和黄土坡一带，是迄今为止北京发现的年代最早的燕国王侯墓葬。这座燕国王侯墓葬区占地5万平方米，共发掘墓葬200余座，车马坑30余座，是北京规模最大的墓葬群。墓葬大致可分为大、中、小三种类型，皆为长方形土坑竖穴墓。大型墓墓圹长、宽都在8米左右，墓室深度达7米以上，有的深达11米。墓道的形制分为长方形土圹木椁墓，一般带有2条或者4条墓道。中型墓墓圹长、宽都在4米左右，墓室为长方形土坑竖穴，墓穴较深，椁室四周一般有二层台，大中型墓葬一般都有车马坑附葬，随葬品多为青铜礼器、兵器、铜工具、玉石器、陶器、漆器等。小型墓墓圹长、宽都在2米左右，墓室为长方形土坑竖穴，墓穴较浅，四周一般有二层台，没有墓道，随葬品多为陶器，或者没有任何随葬品的。

琉璃河燕国王侯墓葬区、昌平白浮泉西周木椁墓和延庆军都山山戎遗址等发掘资料，证明处于西周时期的燕国，厚葬制度已经出现并盛行。

## 二、厚葬的代表——北京的汉代王侯墓葬

北京汉代的墓葬为数不少。西汉政权确立后，北京仍属燕地范围，汉代实行郡县和封国并行的制度，因此决定了北京的王侯高官墓葬的资料较为丰富，最值得提出的是丰台区大葆台的西汉墓和石景山区的老山汉墓。

北京大葆台西汉墓，为大型的土圹木椁墓，墓平面呈"凸"字形，墓坑口大底小形如斗状，上方原有残高9米的封土堆，全墓由墓道、甬道、外回廊、黄肠题凑、内回廊、前室、后室等组成。墓葬的建筑规模极大，建筑形制继承和完善了战国以来采用的不透水胶泥、积沙、积炭和紧密夯土等防护措施，特别是梓宫、便房和黄肠题凑的发现，填补了北京汉墓建筑的空白。梓宫、便房和黄肠题凑是古代统治者使用的最高等级的墓葬制度，梓宫是帝王做棺椁时使用的专用梓木，因此尊称帝王棺椁为"梓宫"；便房位于墓葬的中部，内设宽大的黑漆朱彩坐榻，是帝王的坐席，象征帝王生前起居饮食娱乐的地方；所谓"黄肠"是指堆垒在棺椁外的黄心柏木枋，"题凑"指木枋的头一律向内排列，"黄肠题凑"指西汉帝王陵寝椁室四周用柏木枋堆垒成的框形结构。大葆台西汉墓黄肠题凑用15800余根上等柏木枋围堆成高3米，厚近1米，总长42米的木墙，是北京汉墓保存最为重要的建筑形制。这种建造方式表明，西汉处于中国封建社会第一个鼎盛时期，厚葬之风盛行，上从皇帝，下至王公诸侯，都不惜斥巨资修建陵墓。这种建造形式后来被砖室墓所替代，但这种豪华的建筑形式，为早期封建社会的厚葬习俗提供了丰富的实物资料。

## 三、北京的第一处帝王陵——金陵

北京真正的皇陵出现在金代。陵墓位于房山区西北云峰山下，是北京地区第一处集中埋葬的帝王陵墓群。

金代海陵王完颜亮迁都北京后，决定仿照历代王朝的制度，在都城附近营建皇陵。经过勘陵考察，决定将皇陵建在中都西南大房山中的云峰山下。从贞元三年（1155年）三月开始至十月，在云峰寺旧址修建了一座墓穴，安葬他以前的3个皇帝。第二年又将金朝建国之前的10个祖先灵柩迁来此地安葬，各立称号。经过金代的海陵王、世宗、章宗、卫绍王、宣宗五世60年的营造，共建有17座帝陵，陵区范围达70～80华里，成为北京地区年代最早、规模最大

的帝王陵，比正式列入《世界遗产名录》的明十三陵要早200多年。

金陵以云峰山为主峰，向两翼逐渐延伸开去，占地面积6万多平方米。主陵区在九龙山，主峰高处称为皇陵尖，又叫主龙脉，山势奇秀，有9道山脊顺峰而上，犹如9条巨龙腾云而起，故名九龙山，是一处不可多得的陵寝吉地。陵墓主要分为三个部分：帝陵、妃陵及诸王兆域。金陵的地面建筑和地宫的规模相当宏伟，具有我国北方少数民族的特色。陵区以神道为中轴线，两侧对称布局，由石桥、神道、石踏道、鹊台、东西大殿、陵墙、陵寝等组成。金代的地宫以石椁为帝王陵墓的建筑形式，就地取材，利用当地的石材，作为棺椁建造的材料。石椁既坚固而又保存持久，这种建造陵墓的做法，为以后帝王陵墓开创了先河。金陵墓葬中出土4具石椁，其中，雕龙纹、凤纹的汉白玉石椁，是国内首次发现，为皇室专用的建筑构件和建筑雕饰。

明代末期因帝王迷信风水，将金代皇陵地面建筑全部拆毁，地宫也曾被盗掘。清代女真后裔，对陵墓进行了局部整修，修葺了太祖、世宗二陵。民国以及20世纪六七十年代，山岭遭到严重破坏，地面建筑荡然无存，虽已成遗址，但陵园所存在的宏伟气势，与历代皇陵相比，决不逊色。

## 四、北京的元代墓葬

北京发现的元代墓葬很多，但是元代蒙古人的墓葬不多，特别是蒙古贵族墓葬更为罕见。蒙古贵族墓葬形式与汉族的墓葬形式完全不同，按照习俗，元代帝王的墓葬都采用密葬和薄葬形式，蒙古贵族死后墓地一般被迁回其民族发源地，古称漠北和林等地（今蒙古人民共和国境内），墓葬不砌坟冢，并且将墓葬埋藏地用马牛踏平，不留痕迹。所以至今仍未发现一座元代皇家官宦陵墓。

## 五、气势恢宏的帝王陵墓群——明十三陵

明代是汉族重新在少数民族手中夺回政权的历史时期，建国之初就开始全面恢复古代的礼制，帝陵也一样，恢复和继承了一系列唐宋以来的陵寝制度，如"因山为陵"、帝后同陵、前后各代陵墓集中在同一兆域，并以祖陵为起点，后代陵墓按照昭、穆的顺序分列左右依次排列等，同时，明陵也与前代陵墓产生非常大的变化，并且形成了以方城明楼——宝城宝顶（即以方形的明楼和圆形的陵体）为特色的陵墓建筑群，并将诸陵合用一条公共神道，此外，明代陵寝将唐宋时期地宫分为上、下二宫的形式合而为一，改变了以陵体为中心，四向开门的方形布局，确立了以祾恩殿为中心的长方形的陵区布局，而且这些一直影响到清代的陵墓建造，形成了明清两代的陵墓特色。

明代建造陵墓首先从吴元年（1367年）开始，埋葬朱元璋的父亲。陵址在安徽省凤阳县西南郊，是在其父亲原来坟墓的基础上建造的，至洪武十二年（1379年）完工，初名"英陵"，后改名"皇陵"。此陵的建造受前代影响还是比较大的，但它却是明代皇帝陵墓改革的开端。明代对陵墓的改革在洪武十七年（1384年）营造明"祖陵"的过程中得到了进一步的发展，明代皇帝陵墓真正的成型是位于江苏省南京市钟山南坡明太祖朱元璋的孝陵，陵区内有两道围墙，中轴线上自前而后依次建碑亭、石桥、石像生、棂星门、石桥、陵门、祾恩门、祾恩殿（享殿）、内陵门、石桥、方城明楼和宝城宝顶，孝陵的这种布局方式开创了中国历史上陵墓建设的新规制，并且一直延续使用了500多年。

明成祖迁都北京，自此以后的明代的14位皇帝中，有13位皇帝都葬在了北京昌平区天寿山麓，后人统称为十三陵。十三陵在大约80平方公里的范围内，以一条公共的神道作为进入陵区的总入口，入口处依次建石牌坊、大红门、碑亭、华表、神道柱、石像生及棂星门，陵区内每位皇帝的陵墓占据一座山包，每座陵墓的规模大小不一，但是规制都模仿孝陵，其中明成祖朱棣的长陵规模最大，明思宗朱由检（崇祯皇帝）的陵墓最小。

此外迁都北京之后还有一位皇帝的陵墓没有建在十三陵内，那就是在明代著名的宫廷政变"夺门之变"当中被废黜的景泰皇帝，他被葬在了北京市海淀区，由于此陵最初是

按照亲王礼下葬，后来按照帝陵加以改造，所以规模很小。

## 六、北京的清代王爷墓

　　清代秉承明朝的陵墓制度，建筑形式基本上沿袭了明代的做法。清代墓葬分为皇陵和王爷墓两种。清代兴起于关外满洲，早期建都于盛京，历经数十年的征战，最后定都于北京，出于政权统治的需要，清代帝陵分别修建于河北省的遵化和易县，因此，清代北京所遗存的墓葬基本上都是王爷墓，这是清代陵墓制度的一个突出特点。这些王爷墓虽然规模较皇帝陵寝要小得多，建筑规格降低，但是建筑布局和建筑形式仍旧保留了明代皇家陵寝制度的基本规制。

　　清代与明代亲王的分封制度截然不同，主要区别是，明代亲王可分封外地，领辖一方，死后可以在封地上建造较大规模的坟茔，世代相承。清代皇室没有封地，清代的皇族贵戚，都是仅有爵位、俸禄，生前都住在京城内的官宅之内。因此，墓地只能选在京城郊外。据统计，清代北京共有王爷240位，这些王爷的爵位大小不等，但是，都先后按照一定的规制在京郊建有墓冢。

　　清代等级制度严格，特别是在陵墓制度上等级森严，建造坟茔，绝对不能逾制。从北京现存的王爷墓资料看，王爷墓一般规模不大，较小的坟茔占地约20亩，大型的坟茔占地约200余亩，一般的王爷墓占地大约在百亩左右。建筑布局主次分明，左右对称。墓前挖有环陵墓河道，其上建有神桥，后为左右班房，正中为宫门、享殿，左右建朝房，最后为宝顶封土。有些功勋王爷墓，还建有碑亭、石牌坊等。坟茔的主体建筑为红墙绿琉璃瓦，减皇陵一等。有的王爷墓地附近还建有阳宅，墓地周围广栽松柏，环境肃穆。目前，北京清代王爷墓保存下来的数量不多，以醇亲王墓(七王坟)保存最为完整。

# Tombs in Beijing

Beijing is a cultural capital boasting a history of 3,000 years. Generations of noblemen have lived and been buried here. Especially during the Yuan, Ming and Qing Dynasties in the late of feudal society, Beijing became the political, economic and cultural center of China. With higher political status, Beijing saw the highest rank of mausoleum shifting from that for princes to for emperors, and even super imperial tomb complex. The Jin Tombs in Fangshan District and the Ming Tombs in Changping District are representatives of the noblest tombs in Beijing.

## 1. The Earliest Nobleman Tomb in Beijing: Nobleman Tombs of the Yan State at Liulihe

The earliest nobleman tombs of the Yan State that have been discovered in Beijing so far are at the site of the Shang and Zhou Dynasties, near Dongjialin Village and Huangtupo, Liulihe, Fangshan District, Beijing. The tomb area takes 50,000 square meters consisting of over 200 tombs and more than 30 chariot pits. It is the largest tomb complex in Beijing. The tombs are generally in 3 sizes: large, middle and small. They are all oblong. Large ones are about 8 meters long, 8 meters wide, and 7 meters deep. Some are even 11 meters deep. These tombs mostly adopted oblong earthen shaft containing wooden chamber and generally 2 or 4 tomb passages. Middle-sized tombs are about 4 meters long and 4 meters wide, and the vaults are oblong earthen-shaft graves. There is generally a 2-tier platform around the chamber. Large- and middle-sized tombs typically have chariot pits, and the burial articles are mostly bronze rite wares, weapons, copper tools, jade wares, ceramics, lacquer wares, etc. Small-sized ones are about 2 meters long and 2 meters wide, and the vaults are oblong earthen-shaft graves, relatively shallow. There is generally also a 2-tier platform around the chamber but no tomb passage. The burial articles inside are mostly ceramics, or nothing at all.

## 2. Representatives of Elaborate Tombs: Nobleman Tombs of the Han Dynasty in Beijing

Among lots of tombs of the Han Dynasty in Beijing, the most notable are the tombs of the Western Han Dynasty at Dabaotai in Fengtai District, as well as Laoshan Tombs of the Han Dynasty in Shijingshan District.

The mausoleum of the Western Han Dynasty at Dabaotai is a large earthen-shaft grave containing wooden chamber. It was "凸" shaped in plane, and the tomb had a large top and a small bottom, shaped like a dipper. It consists of tomb passage, corridor, inner and outer ambulatories, Huangchang Ticou (walls of yellow-core cypresses heading inward around the coffin), front chamber, back chamber, etc. The mausoleum had a large construction scale, and its layout inherited and improved protection measures including impervious clay, sand, charcoal and tight rammed loam since the Period of Warring States. The catalpa coffin, front chamber and Huangchang Ticou that were discovered in tombs of the Han Dynasty in Beijing for the first time, are important research materials of history. Huangchang Ticou in the mausoleum of the Western Han Dynasty at Dabaotai was wood walls of 3 meters high, 1 meter thick and totally 42 meters long piled with more than 15 800 first-class cypress poles, as the most important example of construction system among existing tombs of the Han Dynasty in Beijing.

## 3. The First Imperial Mausoleums in Beijing: the Jin Tombs

Actual emperors' mausoleums in Beijing emerged in the Jin Dynasty. Located at the foot of Mount Yunfeng to the northwest of Fangshan District, the mausoleums are the first assembly of imperial tombs in Beijing area.

After migrating the capital to Zhongdu (today's Beijing), Wanyan Liang, the King Hailing, decided to build imperial tomb complex near the capital by imitating the system of the past dynasties. After surveys, he fixed the site at Mount Yunfeng. With the construction by him and other emperors including Shizong, Zhangzong, Weishao and Xuanzong during 60 years, there had built 17 imperial tombs, shaping a tomb area of 35-40 km around. It is the earliest and largest-sized imperial tomb complex in Beijing area, more than 200 years earlier than the Ming Tombs, which has been officially put into World Cultural Heritage Catalogue.

The mausoleums mainly include 3 parts: emperors', queens' and princes' tombs. The ground constructions and underground palaces of the Jin Tombs have very grand sizes. The deity way serves as the axis of the tomb area with 2-side symmetry layout, which consists of stone bridge, deity way, stone passage, magpie platform, eastern and western halls, mausoleum walls, tombs, etc. The underground palaces of the Jin Dynasty adopted stone chamber for imperial burials.

## 4. Tombs of the Yuan Dynasty in Beijing

Lots of mausoleums of the Yuan Dynasty have been discovered in Beijing, but few of them belong to Mongolians, the ruling nationality then, and Mongolian nobleman tombs are especially rare. The form of Mongolian nobleman tombs was totally different from that of the Han Nationality. According to the convention, imperial tombs of the Yuan Dynasty were built intensively and modestly, and the bodies of Mongolian noblemen would be finally returned to be buried in their origin of nationality, ancient Mobei Helin (within today's the People's Republic of Mongolia). The temporary tombs had no ground construction and the Mongolians drove horses and ox on the sites to eliminate any trace. No royal or official mausoleum of the Yuan Dynasty has been discovered yet.

## 5. Grand Imperial Tomb Complex: the Ming Tombs

The Ming Dynasty inherited the system of "building the tomb near the mount", that emperors were buried in the same vault with their queens, and that all tombs assembled in a same zone. The Ming Tombs reformed the mausoleum system of the Tang and Song Dynasties by combining the upper and lower halls and creating the mausoleum complex with square city, memorial shrine, treasure city and tomb being the main constructions.

Located at the southwestern suburbs of Fengyang County in An'hui Province, Yingling is the mausoleum for Zhu Yuanzhang's father. It was constructed at the base of his father's original tomb. Construction of the mausoleum began in the 1st year of the reign of Emperor Hongwu of the Ming Dynasty (1367 A.D.) and was completed in 1380. The tomb plan in the Ming Dynasty started from Xiaoling. Located at the southern slope of the Purple Hills of Nanjing, Xiaoling is the mausoleum of Emperor Taizu, Zhu Yuanzhang. The main buildings, from the front to the rear, are stele pavilion, stone bridge, stone statues, star-worshipping gate, stone bridge, entrance gate, Ling'en gate, Ling'en hall, inner Ling gate, stone bridge, square city, memorial shrine, treasure city and tomb. It's the milestone in the historical development of Chinese mausoleums.

After Emperor Yongle of the Ming Dynasty moved the capital to Beijing, 13 emperors were buried at the foot of Tianshou Mountain in Changping District. Tombs comprised of the mausoleums of the 13 emperors, known as the 13 Ming Tombs. The Ming Tombs cover an area of more than 80 square kilometers. A shared sacred road, common stone archway, great red gate, stele pavilion, ornamental column, stone statues, and star-worshipping gate combine the architectures in the Ming Tombs together. Though the plan for the mausoleums basically continued that of Xiaoling, the scales of them are not the same. Among them, the scale of Changling is the largest, while the tomb of Emperor Chongzhen is the smallest.

Emperor Jingtai was the only emperor who wasn't buried in the Ming Tombs. He was buried in Haidian District. As Jingtailing was repaired and expended from a prince's tomb to an imperial tomb, its scale is very small.

## 6. Prince Tombs of the Qing Dynasty in Beijing

The Qing Dynasty had a very different system of feudal lord from that in the Ming Dynasty. The royal family of the Qing Dynasty had no feudal lands, but rank of nobility, salary and household within the capital. Therefore their tombs can only be built in the suburb of the capital. According to the statistics, there were totally 240 princes of different ranks in Beijing during the Qing Dynasty, and all of them were buried in the suburb in tombs adhering to a certain system.

The Qing Dynasty had a strict class system, especially in the case of the mausoleum, which tolerated no violation of the rank rights. According to the materials about existing prince tombs in Beijing, the size of prince tombs was generally small. Smaller-sized ones occupied about 20 Mu, large ones about 200 Mu, and generally about 100 Mu. The constructions featured clear main-ancillary layout and left-right symmetry. The tombs were surrounded by rivers with stone bridge, left and right wing rooms at the back, gate and sacrificial hall in the middle, courts on the left and the right, and the top of the treasure city. Some tombs for meritorious princes had tablet pavilion, stone memorial archway, etc. The main constructions of the tombs adopted red wall and green glazed tiles, representing the rank just under the imperial tombs. Some prince tombs had Yangzhai, the house for the alive, built nearby, surrounded by a forest of pine and cypress that created a solemn atmosphere. Now, only a few tombs for princes of the Qing Dynasty in Beijing are conserved, and the most complete one is Prince Chun's Tomb (the Grave of the Seventh Prince).

# 十三陵
# The Ming Tombs

明十三陵位于北京市昌平区北部天寿山山麓，明代定都北京后，有13位皇帝埋葬于此，故称十三陵。1961年被公布为全国重点文物保护单位。2003年被联合国教科文组织列入《世界遗产名录》明清皇家陵寝的扩展项目。

明十三陵自明永乐七年(1409年)开始营建，至清代顺治初年，前后长达200余年，形成了体系完整、规模宏大、气势磅礴的陵寝建筑群，是明代帝陵最具代表性的陵墓群，也是我国保存完整、埋葬皇帝最多的帝王墓葬群。

明十三陵陵域面积约为120平方公里。整个陵区四面群山环抱，明堂广大，依山势筑有围墙，陵墓建筑群以长陵为中心，呈环绕之势，掩映在绿树丛林之中。

十三陵诸陵营建情况各不相同，凡是皇帝生前营建的，规模都比较大，例如，长陵、定陵。死后营建的，规模就小，如献陵、景陵、康陵等。思陵，因明末崇祯皇帝是亡国之君，所用陵墓原是贵妃田氏的墓穴，因此，十三陵中属思陵规模最小。

明十三陵陵区的最重要的建筑形制分为祭祀区和埋葬区两部分。地面建筑主要有神道、大红门、神功圣德碑亭、陵园、陵监等以及相关的附属建筑。地下部分称地宫，是帝后棺椁的停放区域。

陵园又称陵宫，是各帝陵地面建筑的集中所在，是祭祀活动的重要专用场所。各陵皆背山而建，方向、规模虽然不同，但地面建筑布局、形制基本一致，特别是主要建筑几乎相同。明十三陵各陵宫殿建筑自成为整体，祭殿在前，寝宫在后。门廊、殿堂、明楼、宝城，排列层次分明，严肃整齐，突出了陵墓建筑的特点。建筑分布以中轴线为主体，随着地势的逐步升高，建筑高低错落有致。陵园祭祀区以墙垣围绕呈长方形，共为三进院落，主要建筑都建在一条中轴线上，两侧有附属建筑陪衬，建筑布局合理完整，与中国传统建筑的布局相吻合。宫门前有无字碑，碑前建石桥与神道相接。第一进院以祾恩门为主要建筑，两侧建有神厨、神库、宰牲亭等附属建筑。第二进院是祭祀区的主体所在，建筑规模和建筑等级都是最高的。主殿以祾恩殿为主，两侧建有配殿相呼应。第三进院由内红门、二柱门组成，后半部有石几筵一座，院内宽敞。地宫区以方城明楼为入口，缭以圆形(或椭圆形)城墙，称之为宝城。宝城内的陵体称为宝顶，地宫即在宝顶之下。这种享殿在前宝城在后的建筑形式，与中国传统宗庙"前庙后寝"同出一源。

大红门
**The Great Red Gate**

大红门是十三陵的总门户。三洞券门，墙体用砖石砌筑，通体皆涂红色，故称大红门。大门的两侧接陵区围墙。门内东侧稍远处原有拂尘殿，又称时陟殿，是专为帝后所建的更衣之所，现已不存。

Located at the foot of Tianshou Mountain to the north of Changping District of Beijing, the Ming Tombs comprised of the mausoleums of the 13 emperors of the Ming Dynasty after it moved its capital to Beijing, known as the 13 Ming Tombs. Listed as a national key relic under special preservation in 1961, as an extension program of the Imperial Tombs of the Ming and Qing Dynasties, the Ming Tombs was listed into World Cultural Heritage Catalogue by UNESCO in 2003.

Construction of the Ming Tombs started in the 7th year of the reign of Emperor Yongle of the Ming Dynasty (1409 A.D.), and ended in the early years of the reign of Emperor Shunzhi of the Qing Dynasty, spanning more than two hundred years. As the representative of mausoleums for emperors in the Ming Dynasty, the Ming Tombs are the largest, the most completed and the most splendid mausoleum complex.

The mausoleum area covers an area of about 120 square kilometers, encircled by mountains. In the central position lies Changling. The other 12 tombs are distributed according to the terrain.

The Ming Tombs have their unique characteristics in their construction. Some large-scale mausoleums were built before emperors'death such as Changling and Dingling, while other small-scale mausoleums were built after emperors' death such as Xianling, Jingling, Kangling etc. As no mausoleum was built before his death for the last Emperor Chongzhen of Ming Dynasty, he had to be buried in the tomb of his imperial concubine surnamed Tian.

Architectures of the Ming Tombs can be divided into the sacrificial area and the burial area. The principal structures on the ground are sacred road, bright red gate, divine merit stele, cemetery and other additional buildings.

Built at the foot of mountains, all the mausoleums have similar architectural layouts and positions on the ground. In particular, the main buildings are almost the same. Each mausoleum has its own sacrificial hall, memorial shrine, treasure city, forming an independent unit. The principal structures stand on the middle axis one higher than the other. The sacrificial area shaped like a rectangle comprises three courtyards one behind another, surrounded by walls. On the middle axis are the main buildings flanked by additional architectures in harmony with traditional Chinese architectural layout. In front of the gate stands a tablet without character. The Ling'en Gate is at the entrance of the first courtyard. There are divine kitchen, divine warehouse and sacrificial animals pavilion in the left and right of the first courtyard respectively. On the second courtyard, the Ling'en Hall where all the important sacrificial ceremonies were held, is the most important architecture of the mausoleum on the ground, flanked by wing halls on both sides. The Memorial shrine is the entrance of the underground palace, surrounded with round walls.

下马碑
The Dismounting Stele

石牌坊
The Stone Archway

石牌坊是整个陵区导入口的标志性建筑，位于陵区的正南面。石牌坊
全部为白石结构，共计五间，宽33.6米、高10.5米，构件上雕有龙、
狮、花卉等精美的图案，精丽宏伟，是北京形体最大、雕刻等级最高
和最精美的石牌坊。

石牌坊（局部）
Section of the Stone Archway

长陵神功圣德碑亭及华表
The Pavilion of the Divine Merit Stele and the
Ornamental Columns in Changling

长陵神功圣德碑
The Divine Merit Stele in Changling

神道石像生
The Stone Statues on the Sacred Road

石像生计有石人12座，分别为文臣、武臣、勋臣。石兽24座，分别
为狮子、獬豸、骆驼、象、马、麒麟。石像生位于神道两侧，相向
成对放置。

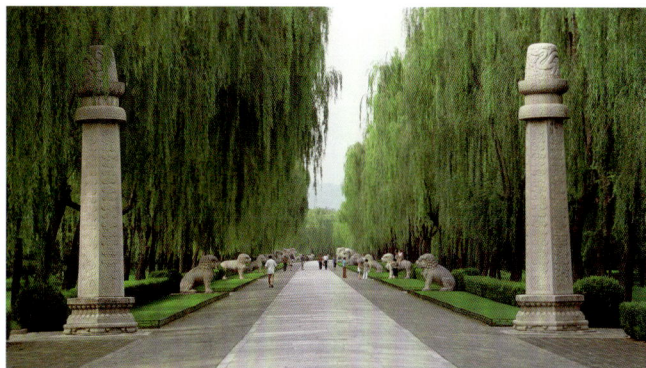

神路
The Sacred Road

神路是十三座陵寝共用的神道，是明代帝陵地面建筑的导引部分。神道南起石牌坊，北至长陵，全长1060米。神道建有石像生、棂星门、五孔桥、七孔桥等。从七孔桥开始，神道由此分支，通向各陵园。

麒麟（局部）
Section of the Kylin

狮
The Lion

獬豸
The Stone Xiezhi Mythical Goat

骆驼
The Camel

象
The Elephant

将军(局部)
Section of the Mililitary Commander

功臣
The Minister of Merit

将军
The Military Commander

龙凤门
The Dragon and Phoenix Gate

长陵祾恩门匾额
The Board Inscribed with "the Ling'
en Gate of Changling"

长陵祾恩门
The Ling'en Gate of Changling

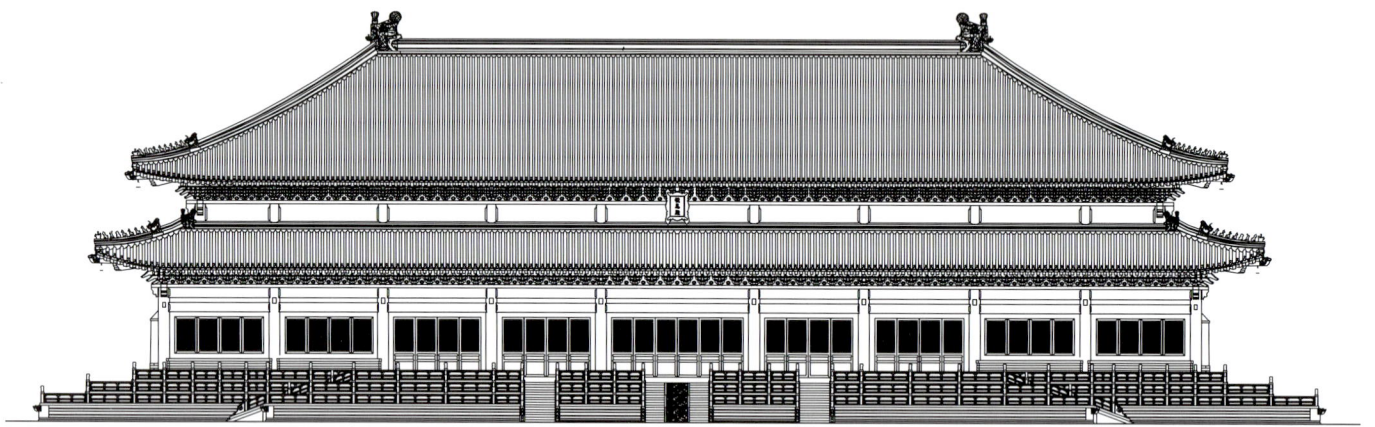

明长陵祾恩殿正立面图
The Elevation of the Ling'en Hall of Changling

长陵祾恩殿御路石雕
The Stone Carvings of the Imperial
Road in the Front of the Ling'en
Hall of Changling

长陵祾恩殿楠木柱
The Corbel Brackets in the
Ling'en Hall of Changling

长陵祾恩殿天花板
The Ceiling of the Ling'en Hall of Changling

长陵石五供
The Stone Five Offerings in Changling

长陵棂星门
The Star-Worshipping Gate in Changling

长陵明楼右侧马道
The Bridle Road to the Right of the Memorial Shrine in Changling

景陵远眺
A Far Sight of Jingling

裕陵
Yuling

永陵祾恩门御路石雕
The Stone Carvings of the Imperial Road in the Front of the Ling'en Gate
of  Yongling

昭陵秋色
Autumn Scenery of Zhaoling

定陵全貌
A Panorama of Dingling

定陵棂星门
The Star-worshipping Gate in Dingling

定陵地下宫殿
The Underground Palace in Dingling

定陵地下宫殿后殿
The Rear Chamber of the
Underground Palace in Dingling

# 景泰陵
# Jingtailing

景泰陵位于北京市海淀区玉泉山北麓的金山口，为明朝第七代皇帝朱祁钰的陵寝，2001年被公布为全国重点文物保护单位。

朱祁钰是明代迁都北京后唯一一位没有葬在十三陵内的皇帝，其原因有深刻的历史背景。明正统十四年(1449年)，明英宗贸然出击入侵的蒙古也先部，发生了"土木之变"。危急时，朱祁钰即位，年号"景泰"。后从蒙古归来的明英宗发动"夺门之变"，朱祁钰被废，不久即死去。英宗不承认他的皇帝身份，不让其葬在十三陵，以亲王礼葬于京西金山口。宪宗朱见深即位后，为朱祁钰平了反，追复了景泰年号，重新营建陵寝。嘉靖时又将其绿琉璃瓦改建为黄琉璃瓦，使之符合帝陵规制，只是规模要小得多了。清乾隆三十四年(1769年)建碑和御碑亭。清后期，皇陵残破将圮毁，中华人民共和国成立后，逐步修整。

景泰陵由于是亲王墓改扩成皇陵的，因此规模上要比同时期的明皇陵小，建筑体量也较小，有些建筑甚至缺少。但从整体看，皇陵该有的主体建筑都已具备，陵区原有享殿、神库、神厨、宰牲亭、内官房和碑亭，碑亭后面建有祾恩殿，最后为宝城。现陵区内保存有御碑亭、祾恩门和宝顶原有的中路部分建筑。御碑亭建于清乾隆三十四年(1769年)，碑亭坐北朝南，面阔一间，重檐歇山顶，顶覆黄色琉璃瓦，木构架绘以旋子彩画。碑亭内有乾隆御笔碑，碑正面刻乾隆皇帝题《明景泰陵文》，背面刻"大明恭俭康复景皇帝之陵"，为乾隆三十四年(1769年)立石。碑亭后为祾恩门，坐北朝南，面阔三间，单檐硬山灰筒瓦屋面，木构架绘以旋子彩画。宝城宝顶位于皇陵的最后，已残毁，现在地面只残存一小部分封土。

御碑亭
The Imperial Tablet Pavilion

御碑亭内彩画
Coloured Paintings of the Imperial Tablet Pavilion

Located at Jinshan Pass of the northern foot of Mount Yuquan in Haidian District, Jingtailing was the mausoleum for the 7th emperor of the Ming Dynasty Zhu Qiyu, which was listed as a national key relic under special preservation in 2001.

Zhu Qiyu was the only emperor who wasn't buried in the Ming Tombs in the Ming Dynasty. In the 14th year of the reign of Emperor Zhengtong (1449 A.D.), Emperor Yingzong led the military campaign against Wala Clan in Mongolia, and was defeated. He was captured and held captive for a year. So, Zhu Qiyu ascended the throne and named his era name as Jingtai. After the Mongols learned that the Ming government had already changed the new emperor, Yingzong was released eventually. He took the chance to recapture the throne and became an emperor again and renamed his era name as Tianshun. After Jingtai's death, Emperor Tianshun denied his rightful honor to be buried in the 13 Ming Tombs but was instead buried at Jinshan Pass to the west of Beijing as a prince.

As Jingtailing was repaired and expended from a prince's tomb to an imperial tomb, its scale was smaller than that of the Ming Tombs at the same time. Originally the principal structures included the Sacrificial Hall, the Divine Kitchen, the Divine Warehouse, the Sacrificial Animals Pavilion, the Imperial Tablet Pavilion, the Ling'en Hall and the Treasure Hill City. Now, the existing buildings are the Imperial Tablet Pavilion, the Ling'en Gate and some buildings on the Tomb. Built in the 34th year of the reign of Emperor Qianlong of the Qing Dynasty (1769 A. D.), the Imperial Tablet Pavilion, which faces south, is 1 bay wide with a double-eaved gable-and-hip roof. The roof is covered with yellow glazed tiles. The wooden trusses are decorated with polychrome paintings. Behind the Tablet Pavilion is the Ling'en Gate, which faces south, is 3 bays wide with a flush gable roof. The roof is covered with tube-shaped tiles.

御碑亭内碑
The Tablet in the Imperial Tablet Pavilion

# 醇亲王墓
# Prince Chun's Tomb

醇亲王墓位于北京西郊北安河西北十余里的妙高峰古香道旁，是清道光皇帝第七子、光绪皇帝的生父醇亲王奕谮的陵墓，也称"七王坟"，是北京著名的王墓之一。1984年被公布为北京市文物保护单位。

醇亲王墓是奕谮在同治年间因病在西山养息时自己选定的，经过前后20余年营造，耗银百万两，造就了七王坟今天的规模。

醇亲王墓坐西朝东，前方后圆，墓地依山顺势，步步升高。主要建筑全部建于东西中轴线上，主要建筑有碑楼、石拱神桥、隆恩门，主殿的两侧建有朝房等附属建筑，陵墓位于中线的最后。墓地有四个宝顶，正中最大的墓冢合葬着醇亲王和福晋，其余的是妻妾的坟茔，南墙外是其夭亡的子女墓。墓园中遍植白皮松，枝繁叶茂，郁郁葱葱。醇亲王墓是非常罕见的阴阳宅布局，墓北侧就是阳宅。阳宅由五进四合院落组成，院内有祠堂、享殿、过厅、走廊、花园等，布局幽静完整。院中有大小建筑数十间，最低层北跨院的一座殿堂内，有一块石碑，记载着修建七王坟的经过。

Located more than 5 kilometers northwest of Bei'anhe to the western suburb of Beijing, the Prince Chun's Tomb is one of the noted princes' tombs in Beijing.It was listed as a Beijing's relic under preservation in 1984.

Yixuan, Prince Chun, was the 7th son of Emperor Daoguang and the father of Emperor Guangxu.

Yixuan was interred in a tomb of princely rank, now popularly known as the "Grave of the Seventh Prince", located at the foot of Miaogao Peak of the Western Mountains. In the Liao and Jin Dynasties, the Xiangshuiyuan Temple, one of the eight temples of the Western Mountains was built there.

The Grave of the Seventh Prince, which faces east, is square in the front and round in the rear. On the east-west axis are the principal structures, which are the archway, the Divine Bridge, the Ling'en Gate, the main hall and the mausoleum.

石拱神桥
The Divine Bridge

碑亭
The Tablet Pavilion

碑亭斗拱
The Corbel Brackets in the Tablet Pavilion

墓地白皮松
The Lacebark Pines in Prince
Chun's Tomb

其他

THE OTHERS

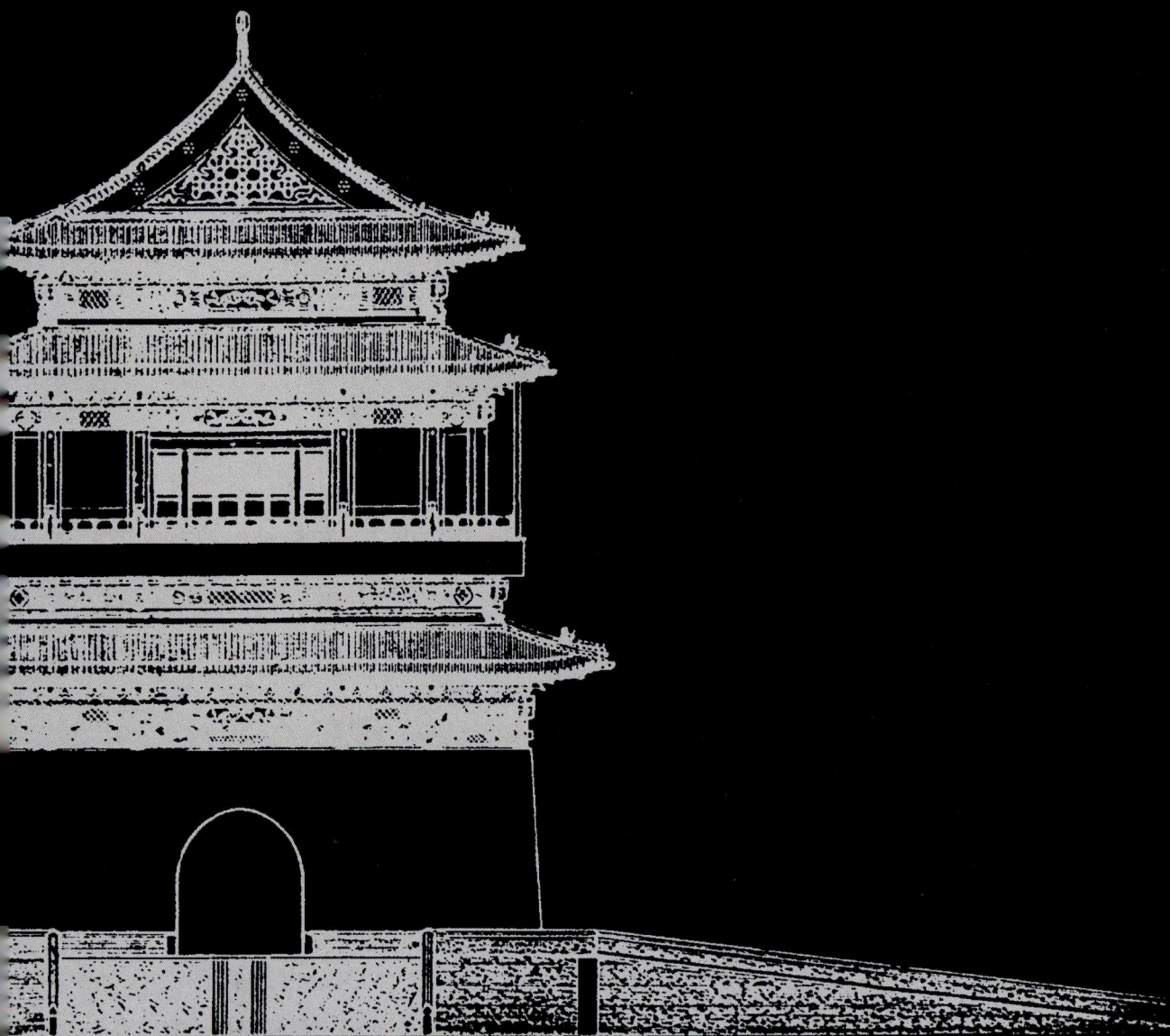

# 北京的其他古代建筑

除了前文所述的八大类古代建筑外，北京遗存下来的古代建筑还有很多，比如各个朝代遗留下来的古桥、明清时期的会馆、位于北京中轴线的南起点的燕墩和北端的钟鼓楼、清代皇帝训练及检阅八旗军的健锐营演武厅、京西古道上的爨底下村古代建筑群等。在这些古建筑中，北京的古桥和会馆数量遗存较多，且独具特色，下面分别加以介绍。

## 一、北京的古桥

古代的北京是一座湖泊众多、河网密布的城市，有河就有桥，因此，历史上北京的古桥不仅数量多，而且形式多样，别具风采。

北京地区保留下来的早期桥梁十分罕见，现存最早的桥梁是金代建造的华北第一长桥——卢沟桥，长达266.5米，用11孔连续的半圆拱构成，成为代表当时建筑水准最高的桥梁建筑，并从此揭开了北京桥梁建造史上的辉煌篇章。元代保存下来的桥梁极少，位于故宫东半部的断虹桥，是北京城唯一保存比较好的元代石桥，从建造形制、建造水平和雕刻艺术多方面，可以看出元代北京建造的桥梁具有很高的水平。明代是北京桥梁建造十分重要的时期，保存到今天的许多著名桥梁建筑都是明代建造的，如朝宗桥、琉璃河大桥、永通桥等。这些桥梁都处于京城周围重要的交通要道上，成为沟通北京与各地的重要通道。清代统治阶级十分重视园林方面的建造，而园林中水网密布，因而园桥的建造成为清代最突出的特点。据统计，仅圆明园就有园桥百余座。除数量多之外，园桥建造的规模和建造艺术也都是首屈一指的，例如颐和园中的十七孔桥，无论是建筑规模、建造艺术、建筑技术和建造环境等都堪称中国园桥之最。因此，可以认为，清代北京皇家园林中的园桥，是北京桥梁发展史上最值得书写的篇章。

北京古桥的种类很多，从形式上划分大概有近10种类型。但从整体上观察，众多类型的桥梁建筑都是从梁桥和拱桥延伸出来的。

北京古代桥梁的建筑艺术，主要表现在三个方面，即造型风格、装饰工艺和附属建筑。综合而论，清代园囿中各式桥梁的建筑艺术更加突出，更具有代表性。

桥梁的造型风格主要体现在曲线柔和、韵律协调。桥梁建筑是建筑与艺术的结合体，是实用与艺术的融合，特别是园囿中专为观赏而特建的各式桥梁，更是独具匠心，纯粹以美术品的要求精工细作，把桥梁所能体现出的美的意境，以实体直觉加以渲染，给人以丰富的想象力和美好的记忆。例如，颐和园西堤所建造的桥梁，桥梁形式虽各具特色，但其艺术共性十分突出。特别是玉带桥的建筑艺术更是美若天成：桥身通体洁白，桥拱高而薄，弧线流畅，形如玉带，与水中倒影重合，恰似一轮圆月，构成一幅动人心魄的美丽画卷。

桥梁装饰工艺主要体现在石构桥梁中，其工艺特征，就是将成熟的雕刻艺术融入于桥梁装饰之中。在以实用性为主的石桥中，有很多建筑构件是为观赏而设置的装饰。其部位大致在人们易于驻足观瞻的地方，如栏板望柱、桥头石兽、拱券券面，甚至桥梁的泊岸等处的雕刻。卢沟桥桥栏上雕刻的石狮，作为护桥的卫士，名闻世界，并成为后代建桥装饰的模式和习尚。

北京现存的不少重要桥梁，往往在桥上或桥头上构建有许多附属建筑物，有碑亭、亭阁建筑、牌坊等。如北京卢沟桥的东西两端，各建有装饰雕刻讲究的碑亭，其中著名的"卢沟晓月"就位于桥梁的东端。颐和园知鱼桥石桥两端的石牌坊，还有北海濠濮涧曲桥北连石牌坊，南接濠濮涧主体建筑。

## 二、北京的会馆

北京历史悠久，建城已3000余年，又是王朝故都，居民和流动人口五方杂处。随着科举的兴旺和工商业的发达，同乡和同业会馆应运而生。北京的会馆以明、清两代建立的最多，清代是会馆发展兴旺的盛期。民国时期也有设立，但数量很少。1949年北京解放后，有的会馆仍发挥着传统的效用，但绝大部分会馆失去了原来的功能。20世纪50年代中期起，400余座会馆逐步变成了民居或工厂厂房。

北京的会馆可概括为两部分：举子会馆与工商会馆。前者包括省、府、州、县等同乡会馆，约占总数的88%，后者为五行八作等手工业和商业会馆，约占总数的12%。这两类会馆的作用虽然都是为便利同乡或同业，但也略有差异，建立的时间、规模也随社会的发展而先后有差别。

北京城市风貌的一大景观是会馆云集，也是京师文化的一大特色。明代商业集中在正阳门外，清初规定内城中只准住旗人，因此明清会馆几乎全在外城。又因中原和南方各省人员进京，最后都要经过涿州，然后向北过卢沟桥，进入广安门，所以绝大多数人便就近停居宣南。正阳门以东则因崇文门为京城总税卡，故多为商人停居之处。在进京人员中，总的来说，商人较少，所以正阳门外西部（宣南）会馆多于东部，而东部则以经商为主的山西籍会馆最多。

会馆建筑按其功能可分为三类：第一类主要是供旅京同乡特别是会考举子居住，就是普通的四合院，但大小悬殊，大的有四五座院，六七十间房屋，如南海会馆；小的一所院落有房不足十间，如广东惠安会馆。第二类是以祭祀、议事、集会为主，兼有少量供达官富商名人居住的小院，占地规模大但房屋不见得很多。著名者有湖广会馆、安徽会馆等。第三类是祭祀议事的场所，不设客房，大部分是行业会馆和会馆附属用房的专祠。如钱业会馆正乙祠（祀正乙玄坛元帅赵公明）。会馆凡冠以某省之名者，多为该省在京任大官（通常为一品大学士）者首倡创立，又多是直接捐出自己府邸为馆舍，所以它们大都保持着大府邸"三轴四部分"的格局。所谓三轴，即中间主院轴和两侧偏院轴；四部分即主院轴为礼仪部分，两偏院轴为居住和书房休闲部分，再有在偏院轴前后的服务供役部分。大府邸改为会馆后，只要把这四部分的功能略加调整，即可满足使用。

会馆之有戏楼由来已久，当初同乡旧友或为京官或为应试举人，每逢朔望节日，相约在会馆聚面，既有酒席陈列，杯盘交错，更有轻歌曼舞，戏曲助兴，所谓"宴"与"乐"常常是分不开的，因此会馆内一般都有戏楼。除近十年陆续修缮完成的湖广会馆戏楼、正乙祠戏楼、安徽会馆戏楼和正在修缮的平阳会馆戏楼，其他的会馆戏楼均已在被拆除之列。

# Other Ancient Architecture in Beijing

Except eight sorts mentioned above, there are also many other ancient buildings existing in Beijing, like ancient bridges built in all dynasties, guild halls in the Ming and Qing Dynasties, the Yan Pier at the southern end of the traditional central axis, the Drum and Bell Towers at the northern end, the Ancient Qing Dynasty Fortress which was a military training compound for emperors of the Qing Dynasty to train and inspect their Manchu troops, the Ancient Village Chuandixia etc. Among all of them, many ancient bridges and guild halls remain, and are of various characteristics. Now, they will be presented respectively.

## 1.Ancient Bridges in Beijing

In Beijing the early bridges still in existence is very rare. The earliest bridge in existence is Lugouqiao Bridge, the longest bridge in northern China that was built in the Jin Dynasty. The 11-multi-arch bridge is 266.5 meters long. The Logouqiao Bridge represented the highest level of bridge architecture at that time, and marked the splendid chapter in the history of bridge construction in Beijing. Very few bridges built in the Yuan Dynasty remained in existence. The Duanhong Bridge, located at the eastern part of the Forbidden City, is a stone bridge built in the Yuan Dynasty and the only one well preserved in Beijing. It shows that the bridge architecture reached a very high level in the Yuan Dynasty, in terms of construction types, construction level and engraving art. The Ming Dynasty was a very important period for the construction of Beijing's bridges. Many famous bridges remaining in existence were built in the Ming Dynasty, such as Chaozong Bridge, Liulihe Bridge, Yongtong Bridge, etc. These bridges all are located in the traffic arteries in Beijing, and become important passage linking Beijing and other regions. The ruling class in the Qing Dynasty attached great importance on the construction of gardens where the water networks were densely covered, therefore,

the construction of bridges in gardens became a most outstanding characteristic. According to statistics, there were more than one hundred bridges in Yuanmingyuan Garden; in addition to the large number of bridges, the bridge construction scale and construction art were unique, for example, the seventeen-arch bridge in the Summer Palace, can be rated as the very best among the Chinese garden bridges in terms of construction art, construction technique, and construction environment. Therefore, it can be regarded that the bridges in Beijing's imperial gardens built in the Qing Dynasty is worthy to describe in the history of the evolvement of Beijing's bridges.

The architectural art in Beijing's ancient bridges are mainly represented in three aspects, i.e. the modeling style, decoration technology, and affiliated architecture.

The modeling style of bridges manifested itself mainly in smooth curve and harmonized rhythm. For example, the bridges constructed in the western bank of the Summer Palace are of various characteristics in the bridge pattern, but their artistic commonness is very extraordinary. In particular, the architectural art in the Yudai Bridge shows natural beauty: the entire bridge body is white, and the bridge arch is high and thin with smooth curve. The bridge takes on a jade belt shape and meets the inverted image in water, just like a full moon.

The decoration technology manifested itself mainly in stone-made bridges, i.e. integrating the mature engraving arts with the bridge decoration. The famous stone lions engraved on the rails of the Lugouqiao Bridge served as guards, and becomes a pattern and common practice that the offspring followed in the construction of bridges.

A number of important bridges existing in Beijing usually have many affiliated buildings, such as stele pavilion, pavilion, and archway. For example, at the eastern and western ends of the Lugouqiao Bridge in Beijing built well-decorated and engraved stele pavilions, including the famous tablet with the inscription "the Moon over the Lugouqiao Bridge at Dawn" standing at the eastern end of the bridge. Other examples are stone archways at both ends of the Zhiyu Bridge in the Summer Palace, and the Haopujianqu Bridge in Beihai Park, which connects the stone archway in the north and the main buildings in the south.

## 2.Guild Halls in Beijing

Beijing, with a long history, went through more than 3,000 years since it became a city. Beijing was an old capital for several dynasties where the residents and floating population lived together. With the prevalence of imperial examinations system and the flourish of the commerce and industry, the guild halls for town fellows and craft brothers came into being. The guild halls in Beijing were mostly built in the Ming and Qing Dynasties, and reached its peak especially in the Qing Dynasties.

The guild halls in Beijing were classified into two parts: guild halls for Juren (first-degree scholars) and guild halls for the commerce and industry. The former included the guild halls for town fellows from the same province, state, prefecture, county and so on, accounting for 88%of the total halls. The latter were handicraft and commerce guild halls, accounting for 12% of the total halls. Therefore, almost all the guild halls in the Ming and Qing Dynasties were established in the outer city. The guild halls at the western side outside Zhengyangmen were more than that at the eastern side, and the guild halls for businessmen most from Shanxi Province were located at the eastern side.

The guild hall constructions were classified into three classes: the first was common quadrangles of various sizes. The large quadrangles had four or five courtyards and less than 70 rooms, such as the Hainan Guild Hall; the small quadrangles had no more than 10 rooms, such as the Hui'an Guild Hall of Guangdong. The second was mainly used for meeting, discussing and sacrificing, only a small number of courtyards which was resided by officials and rich businessmen, had large occupation area but maybe not many rooms, such as the Huguang Guild Hall and Anhui Guild Hall. The third was the place for worship and official business discuss, where there were no guestrooms, and most of the buildings were special temples belonged to the industrial guild halls and affiliated houses, such as the Zhengyi Ancestral Temple. Those guild halls entitled with the names of famous persons in some provinces, were mostly established by the senior officials (generally first-degree great scholars), who contributed their mansions as the guild halls.

Opera theaters had long been a part of the guild halls. When town fellows or old friends became officials or Juren, when the Syzygy Festival came, they gathered in guild halls where they drank and enjoyed songs, dances, and dramas. Since feast is often linked with amusement, opera theaters were usually located in guild halls. Except the opera theaters in Huguang Guild Hall, Zhengyi Ancestral Temple, and Anhui Guild Hall, which were renovated in recent decade, as well as the opera theater in Pingyang Guild Hall under renovation, the opera theaters in other guild halls are included the list for dismantlement.

# 卢沟桥
# Lugou Bridge

卢沟桥位于北京西南郊的永定河上，是北京地区现存最古老的一座联拱石桥,1961年被公布为全国重点文物保护单位。

卢沟桥始建于金大定二十九年(1189年)，迄今已经有800余年的历史。卢沟桥全部用坚固的花岗石建成，全长266.5米，共11孔，石桥为联拱石桥，为适应北方河流夏季洪水泛滥，春季上游冰雪消融伴有"凌汛"的特点，桥墩宽大厚重，桥墩平面呈船形，并且在每个桥墩分水尖安装三角形铁桩，俗称斩冰剑，能够将上游的巨大浮冰撕开撞碎，有效地抗御了冰凌对桥体的撞击，从而保证了桥梁的安全。

卢沟桥的两端共有石望柱270余根，石栏板270余块，并有精美的图案雕刻。望柱顶部雕有高踞的石狮。古代石雕艺术家把石狮的形态、神情刻画得淋漓尽致、千姿百态、神气活现、栩栩如生，其精美的石雕艺术享誉世界。由于桥上雕刻的石狮数目众多，在观赏计数时，稍不留神就会漏过去，于是便有了"卢沟桥的狮子——数不清"的歇后语。直到1984年又一次核查，才最终确定卢沟桥上的石狮共有489只。

卢沟桥的东西两端各设有华表4根，正方形的汉白玉碑亭各一座。其中，东侧碑亭内的"卢沟晓月"碑是金代著名的燕京八景之一。古时卢沟晓月成为当时的重要景观。

桥的东头是宛平县城，这是一座建于明末拱卫京都的拱极城。

卢沟桥
The Lugou Bridge

Spanning the banks of the Yongding River at the southwestern suburb of Beijing, the Lugou Bridge is the oldest of the existing multi-arch stone bridges. It was listed as a national key relic under special preservation in 1961.

First built in the 29th year of the Dading reign of the Jin Dynasty (1189 A.D.), the bridge has a history of more than 800 years. Built of granite, it is 266.5 meters in length and 11 arches. The bridge piers look pointed in shape from the side facing the flow, and square in shape from the reverse side, but as a whole they assume the form of a vessel. The side facing the flow is made with flow-splitting points. And the points are capped with triangular iron poles, popularly called "Zhanbing Swords", in order to meet ice head-on with the points when the ice in the river begins to melt in the spring time and protect the piers.

On both sides of the bridge are 279 baluster columns and 279 balustrade panels, carved with fine patterns. On top of each baluster column, stone lions of different sizes and shapes are carved. These lions look vividly. Sitting, sleeping, rising and falling, they are full of changes. A saying goes like this "There are too many stone lions to be counted on the Lugouqiao Bridge". According to the surveying in 1984, the total sum is 489 lions.

At the each end of the bridge there are 4 ornamental columns and a square white marble tablet pavilion. The tablet with the inscription "the Moon over the Lugou Bridge at Dawn" was one of the eight scenic spots of Yanjing in the Jin Dynasty.

石板路上的车辙印
The Road on the Lugou Bridge

卢沟桥栏板柱头石狮
Lions on Top of Each Baluster Column

石狮
The Stone Lion

石狮
The Stone Lion

石狮
The Stone Lion

石狮
The Stone Lion

石狮
The Stone Lion

石狮
The Stone Lion

石狮
The Stone Lion

石象
The Stone Elephant

斩冰剑
Zhanbing Swords

桥栏外侧的云纹装饰
The Cloud Design on
the Balustracde

镇水兽
The Stone Flood-
Subduing Beast

卢沟晓月碑
The Tablet with the
Inscription "the Moon over
the Lugou Bridge at Dawn"

# 永通桥及石道碑
## Yongtong Bridge

永通桥位于北京市朝阳区，横跨通惠河上。因距通县西8里，俗称八里桥，是京城东面最重要的桥梁。1984年被公布为北京市文物保护单位。

元代至元二十七年(1290年)，为解决从运河运来的南粮入京问题，忽必烈下诏修通惠河运粮进京。永通桥便是通惠河上建造的唯一的一座大型石拱桥，史称它为"陆运京储之通道"。石桥的前身是一座木桥，因通惠河坡度较大，河水湍急，常将这里原建的木桥冲毁，影响交通，因此有内宫太监李德奏于明英宗，建议于此地建石桥，英宗准奏，正统十一年(1446年)十二月竣工，英宗赐名"永通桥"。石桥的建成，不仅解决了交通，控制了洪水，还为古老的通惠河增加了一个美丽壮观的景点——"长桥映月"。

永通桥是一座三孔石拱桥，长60.2米，桥宽12.26米，石桥的建筑造型特殊，中孔特高，高达8.5米，宽6.7米，两侧孔仅高4.31米。主侧孔相差悬殊，这种构造是专为漕运的需要设计的。通惠河运粮船多为帆船，如建造普通形式拱桥，势必阻碍漕船的航行，为此古代工匠将桥的中孔建造得相当高耸，漕船可直出直入，圆满地解决了这一难题，所谓"八里桥不落桅"正是指此。桥上两侧护有青石栏，汉白玉望柱上有各具形态的石雕狮子，桥梁东西泊岸上雕镇水兽4只，作伏踞状，造型奇特，形象生动。

永通桥地处交通要道，是京东入城咽喉，战略位置极为重要，历史上这里曾进行过两次大规模的中外战争。光绪二十六年(1900年)，义和团曾与八国联军激战于此。桥东有雍正十二年(1734年)"御制通州石道碑"，记载了自京师至通州修筑石道的情况，史料价值极为重要。

永通桥
The Yongtong Bridge

Spanning the banks of the Tonghui River in Tongzhou District of Beijing, the Yongtong Bridge is commonly known as Bali Bridge for the reason that it is located 4 kilometers to the west of the county. The bridge is the most important bridges in the east of Beijing and was listed as a Beijing's relic under preservation in 1984.

In the 27th year of the Zhiyuan reign of the Yuan Dynasty (1290 A.D.), Kublai Khan issued an imperial edict to dredged Tonghui River to solve foodstuff transport. The Yongtong Bridge is the unique great stone arch bridge built above the Tonghui River. Wooden bridges used to be built in this place. Because they were damaged by rushing water frequently, Li De, a eunuch of the Ming court, suggested Emperor Yingzong to build a stone bridge here. Construction of the bridge was completed in the 11th year of the reign of Emperor Zhengtong (1446 A.D.). Emperor Yingzong named it as Yongtong Bridge.

The bridge is 60.2 meters in length, 12.26 meters in width and 3 arches. The arch in the middle is 8.5 meters in height and 6.7 meters in width, flanked by two 4.31-meter-high arches. On both sides of the bridge are balustrades and white marble baluster columns which are carved with stone lions of different sizes and shapes.

桥头石雕
Stone Carvings at the End of the Bridge

# 琉璃河大桥
# Liulihe Bridge

琉璃河大桥位于北京市房山区琉璃河镇北的京石公路上，是房山区境内最大的联拱石桥，其规模仅次于北京著名的卢沟桥。1984年被公布为北京市文物保护单位。

琉璃河大桥初建于明嘉靖十八年(1539年)，历时7年。明嘉靖四十年(1561年)又拨帑银修筑路堤并与石桥相连接。彻底解决了琉璃河的水患，从修桥到路堤建成，前后20余年，南北交通要道天堑变通途。

琉璃河石桥呈南北走向，横跨琉璃河上，桥体全部用巨大的石块砌筑，规模宏大，全长170米、宽11米、高8米左右。桥体按河水的实际流量的需要而设计，共九孔，中孔最大，两侧桥孔依次减小。中孔拱券券脸石正中雕刻有镇水兽纹饰，桥上建有实心栏板和望柱，其上均雕有海棠线等纹饰，雕饰古朴简洁。

琉璃河石桥的两端原建有"玄恩"、"咸济"两座牌坊，桥北建神祠，现已无存。琉璃河桥及其周围的美景美色，古代将其列为著名的"良乡八景"之第一景——"燕谷长桥"。

Located on Jingshi Highway to the north of Liulihe Township in Fangshan District, the Liulihe Bridge is the largest multi-arch stone bridge in Fangshan District. The scale of it is only less than that of the noted Lugou Bridge. It was listed as a Beijing's relic under preservation in 1984.

Construction of the bridge began in the 18th year of the reign of Emperor Jiajing of the Ming Dynasty (1539 A.D.). It was completed 7 years later. Emperor Jiajing spent taels of silver on constructing banks to connect them with the bridge. So, the floods of Liulihe River were thoroughly solved. It took more than 20 years to construct the bridge and the banks.

Spanning over the Liulihe River from the north to the south, the grand and magnificent Liulihe Bridge was built of huge blocks. It is 170 meters in length, 11 meters in width, about 8 meters in height and 9 arches. On the bridge are solid balustrade panels and baluster columns carved with fine patterns. It was one of the eight scenic spots of Liangxiang Village.

琉璃河大桥
The Liulihe Bridge

# 万宁桥
# Wanning Bridge

万宁桥位于北京市西城区地安门北侧，又名海子桥、后门桥，是北京城中轴线末端重要的古代桥梁。1984年被公布为北京市文物保护单位。

元代定都北京以后，在古积水潭这一串弓形湖泊的最东部，即今什刹海前海的东岸，画了一条切线作为全城的中轴线。中轴线的起点，正是万宁桥的所在地。元代至元年间(1264－1294年)，在著名水利学家郭守敬的指导下，修建了通惠河，由南方沿大运河北上的漕船经通惠河可径直驶入大都城内的积水潭。万宁桥为积水潭的入口，漕船要进入积水潭，必须从桥下经过。因而，万宁桥不仅是北京漕运的重要遗址，也是北京城起源的重要标志。

万宁桥始建于元代，原桥为木桥，因地处闹市，交通流量过多，运粮船碰撞的原因，木桥多次被毁，后改为石桥。现存的古桥是明代时期建造的。石桥为单孔拱券，券脸石正中雕刻有镇水兽纹饰，怒目而视，形象极为生动。桥梁的东西泊岸上雕镇水兽4只，作伏踞状，造型奇特。桥两端建石护栏，栏板雕饰宝瓶图案，望柱雕莲花头。

The Wanning Bridge is located at the north of An'dingmen in Xicheng District and is also called Haizi Bridge or Houmen Bridge. As an important ancient bridge in the northern end of the central axis, it was listed as a Beijing's relic under preservation in 1984.

First built during the Yuan Dynasty, the bridge was a wooden bridge. As located at the downtown area, it was destroyed by much traffic and the collision of the boats. Later, it was rebuilt as a stone structure. The existing bridge was built during the Ming Dynasty. It is a single arch bridge. The arch is covered with a flood-subduing best with glaring eyes looking vivid. On both sides of the bridge are baluster columns and balustrade panels, carved with fine patterns.

万宁桥
The Wanning Bridge

# 朝宗桥
# Chaozong Bridge

朝宗桥位于北京市昌平区沙河镇北，是京城北部最大的联拱石桥，1984年被公布为北京市文物保护单位。

朝宗桥始建于明代万历年间(1573－1620年)，是通往京北的咽喉，它北控居庸、白羊，东扼古北关口，明军出征扫北，军事地位极为重要。同时也是明代统治阶级谒陵的重要通道。

朝宗桥全部用巨大坚固的花岗石砌筑而成，全长130米，共7孔，桥体从河水的实际流量的需要而设计，中孔最大，两侧桥孔依次减小。中孔拱券券脸石正中雕刻有镇水兽纹饰，怒目而视，该兽是传说中的龙生九子之一图形。桥栏雕饰古朴简洁，给人以雄浑之感觉。朝宗桥的北端东侧，矗立有一座高大的汉白玉螭首方碑，建于万历四年(1576年)，通高4.08米，石碑的阴阳两面雕刻内容相同。碑额篆书"大明"，碑身线刻双钩"朝宗桥"，馆阁体，苍劲有力，丰满圆润。

由于朝宗桥具有谒陵、北巡的重要性，因此，明朝在桥南侧建有皇帝行宫——巩华城，城内设置衙署官吏，并派重兵屯集，扼守京城北路，成为京北拱卫京城的战略要地。

朝宗桥碑
The Tablet of Chaozong Bridge

朝宗桥镇水兽
The Stone Flood-Subduing Beast of Chaozong Bridge

Located to the north of Shahe Township in Changping District, the Chaozong Bridge is the largest multi-arch stone bridge in the north of Beijing. It was listed as a Beijing's relic under preservation in 1984.

First built during the reign of Emperor Wanli of the Ming Dynasty (1573−1620), the bridge used to be a main pass to the north of Beijing. Because of the importance of the bridge, it had been the place that the military fight against for.

The Chaozong Bridge built of huge stones is 130 meters in length and 7 arches. The arch in the middle is carved with a flood-subduing best with glaring eyes, which is one of the nine sons of the dragon in legendary. At the northern end of the bridge stands a 4.08-meter-high square tablet made of white marbles, which was built in the 4th year of the reign of Emperor Wanli (1576 A.D.).

朝宗桥桥体
The Chaozong Bridge

# 湖广会馆
# Huguang Guild Hall

湖广会馆位于北京市宣武区虎坊路3、5号，是湖南、湖北两省的会馆。1984年被公布为北京市文物保护单位。

该处在清乾隆时为张惟寅、王杰、刘权之等官员府邸，嘉庆十二年(1807年)捐为会馆。道光十年(1830年)改修，增建戏楼，扩建文昌阁。道光二十九年(1849年)又重修，增添花园。光绪十八年至二十二年(1892－1896年)再次大修。

据1927年《湖广会馆全图》标注，会馆原占地东西48.77～53.34米，南北82.3～92.66米。1976年拓宽骡马市大街，拆去北部，现状南北64.24米，东西42.8米。湖广会馆分为东、中、西三路。原大门为一木栅栏门，现已无存。现状沿东边巷道至二门垂花门，进门后为会馆东路前院，有五檩倒座房三间；通过游廊至中院，有五檩带前廊东房六间；再北为东路主院，原有多座建筑，现只存三间南房，五檩北向加前廊。中路的主体是戏楼(又名罩棚)，楼后以平顶游廊围成庭院，院中即著名的"子午井"；正面为文昌阁，阁后两边有爬山廊，上至"风雨怀人馆"；再北即新添宝善堂。西院建筑经多次改建，只有楚畹堂尚是原物，堂面阔三间，三卷勾连搭，共十一步架，其前廊为四檩卷棚顶，较为特殊。此堂当年装修雅洁，是两湖名流宴会吟咏之地。

Located at 3 and 5 Hufanglu in Xuanwu District of Beijing, the Huguang Guild Hall used to be a guild hall of Hunan and Hubei Provinces and was listed as a Beijing's relic under preservation in 1984.

This place was the officials' residence during the reign of Emperor Qianlong of the Qing Dynasty. In the 12th year of the reign of Emperor Jiaqing (1807 A.D.), it served as a guild hall. The opera theater was built and the Wenchang Hall was enlarged in the 10th year of the reign of Emperor Daoguang (1830 A.D.). The garden was added in 1849.

According to *The Whole View of Huguang Guild Hall*, the guild hall originally was 48.77-53.34 meters from the east to the west, and 82.3-92.66 meters from the north to the south. In 1976, its northern part was dismantled for widening the Luomashi Street. Now, it is 64.24 meters from the north to the south, and 42.8 meters from the east to the west. The guild hall is divided into the middle, the eastern and the western axes. On the eastern axis, the drooping flowers gate stands in front of the first courtyard, where lies the reversibly-set room with 3 bays wide and five purlins. In the second courtyard, the eastern wing room is 6 bays wide with five purlins and the front corridor. In the third courtyard, there is the southern room with 3 bays wide. On the middle axis, there are the opera theater, the Ziwu Well, the Wenchang Hall and the Baoshan Hall.

戏楼院门
The Gate of Opera Theater

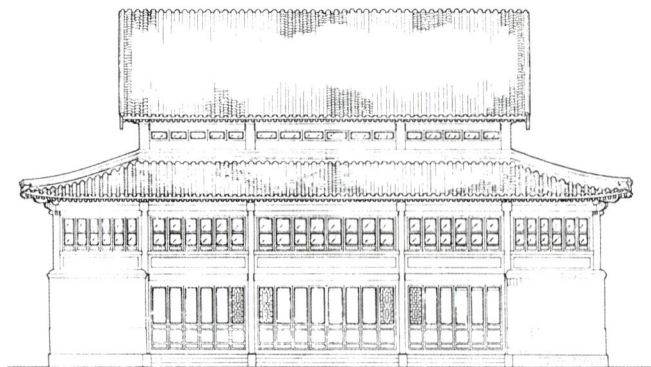

北立面图
The Elevation

垂花门
The Drooping Flowers Gate

会馆游廊
Corridor of the Guild Hall

戏楼
The Opera Theater

湖广会馆的主体建筑为戏楼，面阔五间，当心间即舞台柱间宽度，达5.68米，进深七间，二层楼、东、西、北三面为楼座，南面为舞台；后台五间，高达两层，后再接单坡房五间为扮戏房。戏楼为抬梁式木结构，双卷重檐悬山顶，仰合瓦屋面。上檐双卷高跨为十檩，低跨为六檩，十一架大梁长达11.36米，在北京民间建筑中极为罕见；下檐为楼座屋顶、单坡四檩、外设木板槛墙槛窗；再下为砖砌墙身，开方窗。北面开隔扇门三樘进入游廊，东面开板门两樘，西面突出两间为场面(乐队)使用。楼内原无天花板，梁架和四周走马板、楼座挂檐板及栏杆均有苏式彩画。

# 安徽会馆
# An'hui Guild Hall

安徽会馆位于北京市宣武区后孙公园胡同17、19、23、25、27号，现状范围东西56米，南北74米，为京师最著名的会馆之一。2006年被公布为全国重点文物保护单位。

孙公园原是明末清初著名学者孙承泽的寓所。清末权臣李鸿章及其兄李瀚章（湖广总督），为扩充淮军集团的势力，与淮军诸将共捐万金，于同治七年至同治十年（1868－1871年）在此处建造安徽会馆。1998－2000年对戏楼、文聚堂等进行了全面修缮。

会馆分为中、东、西三路庭院，每路皆为四进。各路庭院间隔以夹道，最北部为一座大型园林。中路为节日聚会、议事、酬神演戏的场所，主体建筑为文聚堂和戏楼。东路为乡贤祠，有思敬堂、奎光阁，东夹道设有箭亭。西路为接待居住用房，隔壁为泉郡会馆。花园面积约两亩余，原有假山亭阁，池塘小桥，现仅存一座"碧玲珑馆"。整组建筑除花园已无存外，基本格局保存尚好，只东路建筑残破拆改严重。

戏楼
The Opera Theater

戏楼是中路规模最大的建筑，面阔五间，双卷勾连搭悬山顶。前为五檩，后为八檩，合瓦顶屋面。戏台在南面，后接扮戏房五间。其余三面为楼座，外观为重檐式。戏楼北部为过厅五间，五檩，前出廊。再后即为"碧玲珑馆"，二层建筑面阔五间，六檩悬山顶。梁架为原物，装修已改。

Located at No.17.19.23.25 and 27 Housungongyuan Hutong in Xuanwu District, the An'hui Guild Hall is 56 meters from the east to the west and 74 meters from the north to the south. As one of the most noted guild halls, it was listed as a national key relic under listed preservation in 2006.

The An'hui Guild Hall was proposed by Li Hongzhang, a famous minister of the last Qing Dynasty, and his brother Li Hanzhang and supported by generals of Huaijun. It was built based on the former address of Sun Chengze's Residence, a famous scholar of the last Ming Dynasty and the early Qing Dynasty in the 7th year of the reign of Emperor Tongzhi (1868 A.D.) and was completed in 1871. The opera theater and Wenju Hall were completely repaired and renovated from 1998 to 2000.

The guild hall is divided into the middle, the eastern and the western axes, each of which comprises four courtyards one behind another. On the northern end stands a large-scale garden. The middle axis was used as a place of meeting, discussing, sacrificing and acting on festival. The main buildings are the opera theater and the Wenju Hall.

碧玲珑馆
The Bilinglong Storied Building

# 正乙祠
# Zhengyi Ancestral Temple

正乙祠位于北京市宣武区前门西河沿街281号，2001年被公布为北京市文物保护单位。

该处明代为寺庙，清康熙六年（1667年）由浙江绍兴银号商人集资，利用古寺址创立祠堂会馆。1710－1712年，修正乙祠，内设戏楼供"燕乐"，大堂供集会，有《正乙祠碑记》。乾隆、同治和民国年间进行过重修或修缮。1989年作局部加固。1994年民营企业家王宇鸣出资抢险修缮正乙祠戏楼，恢复古戏楼面貌，1995年再现演出盛况。

正乙祠坐南朝北，保护范围以现存院墙为界，南北40米，东西30米。临街为九间倒座北房，进深五檩加前廊。正中一间辟为入口，系广亮大门，尺度与一般住宅相似。院内偏东有南房三间，五檩加前廊硬山式；戏楼位于会馆长方形院落的南部，坐南朝北，卷棚歇山顶，木结构，面积315平方米。

戏楼
The Opera Theater

戏楼正中罩棚（即池座）东西面阔三间，南北进深十二檩，二层，上加披檐。戏台在南面，上下二层，伸出式舞台；一层戏台正方形，台基高0.95米，约6米见方，四角立柱。

台顶设木雕花罩，侧面有架空木梯通二层楼座。台下中心为池座，约70平方米。东、西、北三面为二层看楼，进深各2.2米；看楼的楣枋有木雕牡丹纹饰，正面看楼护栏下雕有5条行龙；室内梁架露明，原有彩画。楼座设"万"字花板栏杆，雕花木挂檐板。楼座向外满开槛窗。舞台后（南）部扮戏房六间。戏楼前（北）部正厅五间，正中三间为厅，两侧稍间为戏楼入口。整个戏楼可容纳200余人。正乙祠戏楼布局紧凑，工艺讲究，在会馆戏楼中别具一格。

Located at No.281 Xiheyan Street in Xuanwu District, the Zhengyi Ancestral Temple was listed as a Beijing's relic under preservation in 2001.

A temple was built in this place in the Ming Dynasty. In the 6th year of the reign of Emperor Kangxi of the Qing Dynasty (1667 A.D.), merchants raised money to found a guild hall here. It became the place where they gathered to offer sacrifices to gods. Restoration of the Zhengyi Ancestral Temple started in 1710 and was completed two years later. It was repaired and renovated during the Qing Dynasty and the Republic of China.

The Zhengyi Ancestral Temple, which faces north, is 40 meters from the north to the south, and 30 meters from the east to the west. The northern Reversibly-set room is 9 bays wide and 5 bays deep with the front corridor, in the center of which stands the Guangliang Gate. The opera theater at the south, which faces north, covers an area of 315 square meters.

# 北京鼓楼、钟楼
## Drum Tower and Bell Tower of Beijing

北京鼓楼、钟楼位于北京城南北中轴线的最北端，是元、明、清三代都城的报时中心，也是研究明、清两代建筑形制、建筑构造、建筑艺术和冶炼、铸造、施工等多方面的珍贵资料。1996年被公布为全国重点文物保护单位。

鼓楼始建于元至元九年（1272年），原名"齐政楼"，后毁于火。元成宗大德元年（1297年）重建，之后再度焚毁。明永乐十八年（1420年）再次重建。明嘉靖十八年（1539年）鼓楼遭雷击被毁，第三次重建。清嘉庆、光绪年间都曾对鼓楼进行过不同程度的修缮。1900年，"八国联军"入侵京师时，楼上的文物遭到严重的破坏，建筑幸免于毁。民国十三年（1924年），鼓楼易名为"明耻楼"，在楼上陈列了一些图片和模型，向人们展示八国联军侵略北京的情形，以示不忘国耻，次年恢复原称"齐政楼"。

钟楼原址为元大都大天寿万宁寺之中心阁，始建于元至元九年（1272年），后毁于火。明永乐十八年（1420年）在重建宫室的同时重建了钟楼，尔后再度烧毁。清乾隆十年（1745年）奉旨再次重建，并全部改为砖石无梁拱券式结构，以后又曾多次进行修缮。1976年，唐山大地震波及北京致使钟楼受损，1984年、1986年国家两度拨专款对钟楼进行全面的修缮。

钟、鼓楼均坐北朝南，前后纵置，鼓楼在前，钟楼在后。

The Drum and Bell Towers are situated at the northern end of the central axis of Beijing City. They were the places of striking the hours and are also the treasury for researching system, structure and art of architecture as well as smelting, foundry and construction in the Ming and Qing Dynasties. It was listed as a national key relic under special preservation in 1996.

The Drum Tower was first built in the 9th year of the Zhiyuan reign of the Yuan Dynasty (1272 A.D.), originally known as Qizhenglou (the Tower of Orderly Administration), which was destroyed by fire afterwards. It had undergone three fires and three reconstructions. In the 13th year of the Republic of China (1924 A.D.), the Drum Tower was given another name, Mingchilou (the Remembering Humiliations Tower), which displayed several photos and models, showing people how the Allied Forces of Eight Imperialist Powers invaded Beijing, in order to arouse the people's patriotic enthusiasm. In 1984, the Drum Tower was thoroughly renovated once again.

Construction of the Bell Tower began in 1272, built on the basis of the main hall of the former Wanning Temple of the Yuan Dynasty. It was destroyed by fire. In 1420, it was rebuilt, then burnt down again. In 1745, Emperor Qinglong ordered the Bell Tower to be rebuilt with bricks and stones.

Facing south the Bell Tower stands on the north of the Drum Tower.

位于北京城南北中轴线最北端的鼓楼和钟楼
The Drum and Bell Towers at the Northern End
of the Central Axis of Beijing

鼓楼
The Drum Tower

鼓楼为三重檐歇山木结构建筑，顶覆绿剪边灰筒瓦，通高46.7米，总占地面积6857平方米，建筑面积2736平方米。鼓楼分上下两个功能层和中间的一个结构暗层。楼底层为无梁拱券式砖石结构，共有券门八座，其中南北各三座，东西各一座。东北隅设登楼小券门，内设登楼石阶梯69级直通二层。二层面阔五间，四周有回廊，宽约1.3米，平座周围建木栏杆，四角支撑有擎檐柱。平座下悬木挂檐板，如意头贴金彩画。檐下及平座下均施以斗拱，木构架绘有旋子彩画。整座建筑坐落在高约4米、四面呈坡道形的砖石台基上，台基宽56米，进深33米，南北有砖砌阶梯，东西为礓磋坡道，是京城体量最大的砖木建筑之一。

鼓楼二层原有更鼓二十五面，包括主鼓一面，群鼓二十四面，代表一年和二十四节气，作为鼓楼定更报时的器具，现仅存一面主鼓，并且损毁严重。该鼓高2.22米，腰径1.71米，鼓面直径1.40米。此外，二层原置铜壶滴漏一座，为古代计时器物，但早已遗失无考。当时，钟、鼓二楼击鼓撞钟均以此为度，至清代改用时辰香计时后，滴壶计时法方废止。

2001年将仿制的一面主鼓，二十四面群鼓置于鼓楼二层陈列，并于每天定时进行击鼓表演。2005年根据史料记载，仿制了铜壶滴漏在鼓楼二层陈列、演示。至此鼓楼的计时器具——铜壶滴漏，定时器具——二十五面更鼓在二层组合展出。

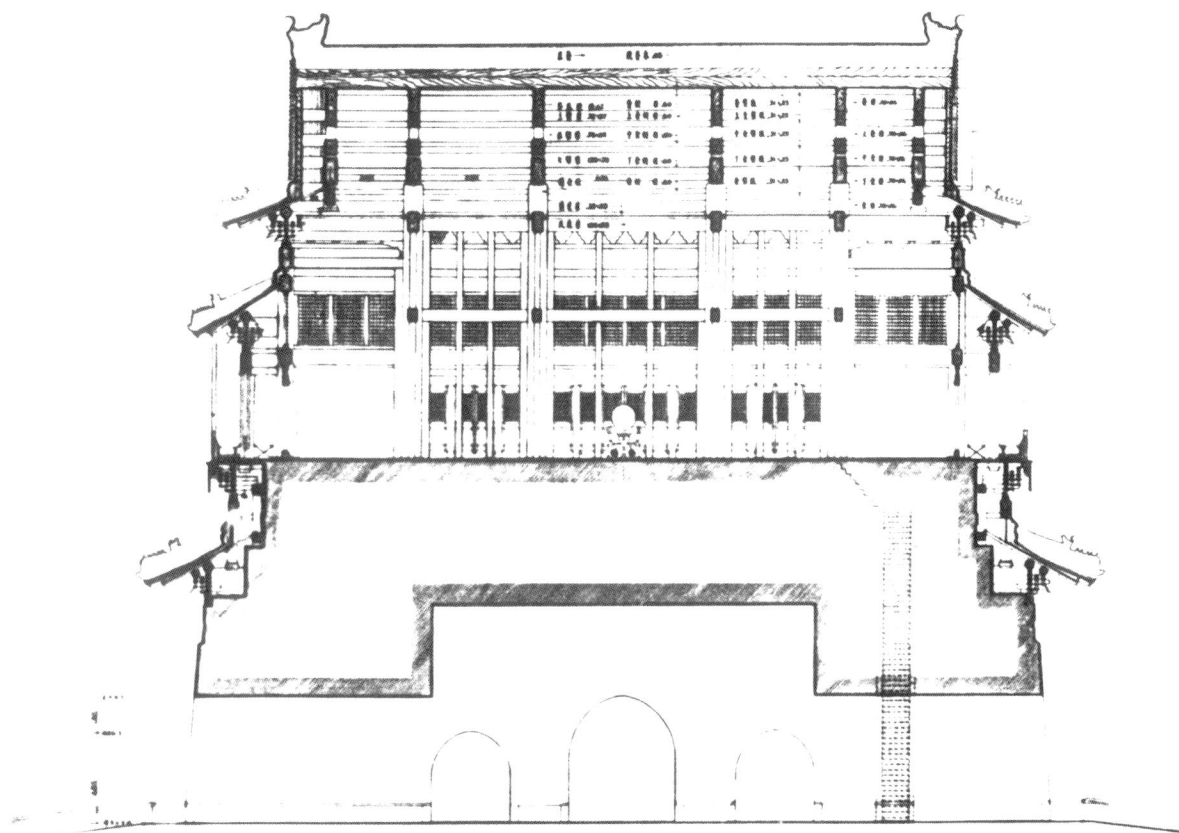

鼓楼南侧立面图
The Elevation of the Drum Tower

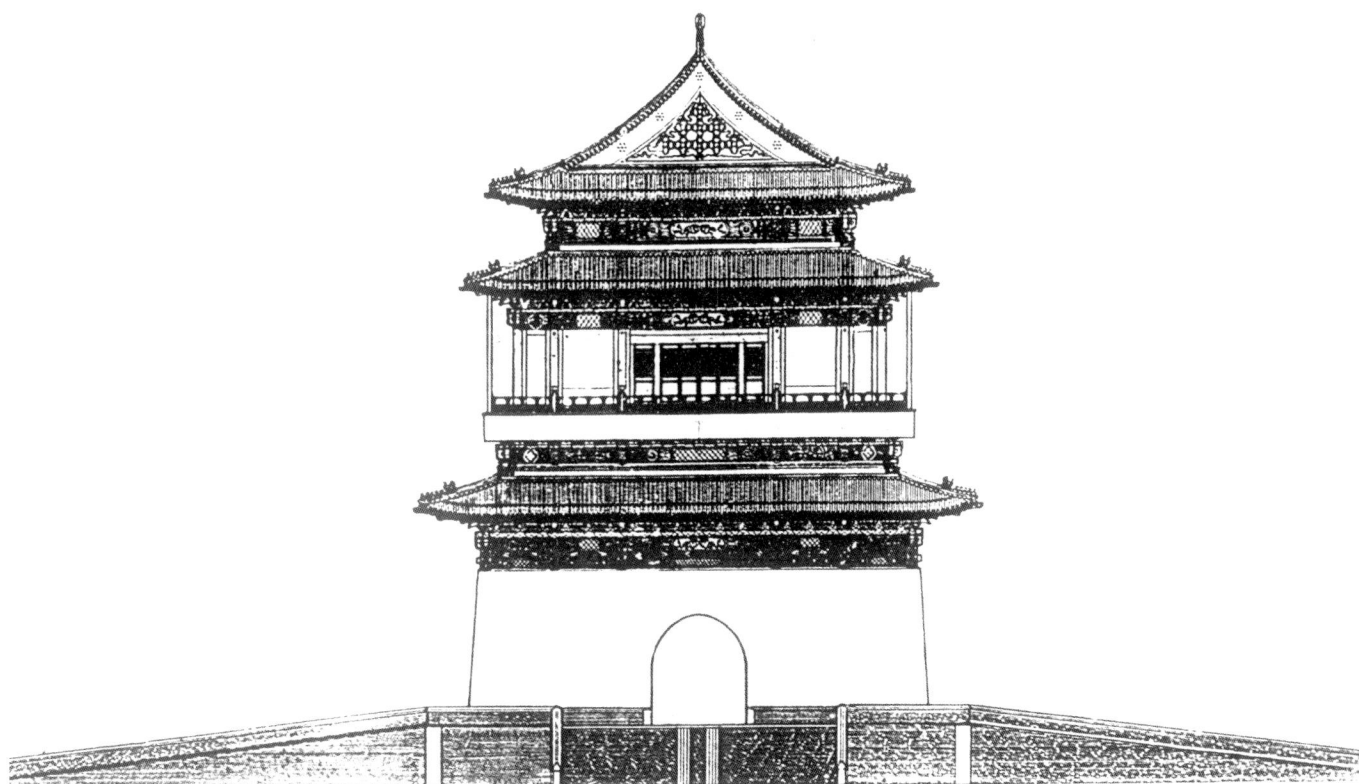

鼓楼西侧立面图
The Elevation of the Drum Tower

钟楼
The Bell Tower

钟楼位于鼓楼北侧，坐北朝南，无梁拱券式砖石结构，通高47.9米，分上下两层，占地面积5740平方米。一层基座为城砖砌筑而成，四面各有拱券式大门一座，上部建有雉堞，内部呈十字券结构。钟楼东北隅有旁门，内有75级石阶直通二层。基座之上为汉白玉须弥座，束腰雕有精美蔓草纹饰，四周环以汉白玉石护栏，石榴柱头，三幅云净瓶镂空栏板，南北各出八级台阶，单侧扶手栏杆，可供上下，须弥座上建造钟楼。钟楼面阔三间，重檐歇山顶，顶覆绿剪边黑琉璃瓦。钟楼四面明间开有拱券门，次间为三交六碗仿木石券窗，边框上雕有铺首。两侧山花均为琉璃砖拼成的金钱绶带图案。建筑檐下石构件绘以旋子彩画。

钟楼原有明永乐年间铸造的铁钟一口，置放于楼外平地上，后由古钟博物馆收藏。现钟楼内正中立有一八角形木框架，上面悬挂着一口铸有"大明永乐年月吉日制"印记的报时铜钟，重约63吨，堪称中国古钟之最。

钟楼正南为面阔三间的大门，明间门内正中立乾隆十二年(1747年)御制碑，乾隆撰文，户部尚书梁诗正书写，碑阴《京兆通俗教育馆记》，刻于1925年。

# 燕墩
# Yan Pier

　　燕墩俗称"烟墩"，位于北京市崇文区西南部，永定门外大街31号。由于地处传统中轴线的南端，因此燕墩是北京城南中轴线的标志性景观。1984年被公布为北京市文物保护单位。

　　燕墩始建于元代，为北京五镇之一的南方之镇，它是历代皇帝用以祈求皇权永固的场所。因南方在"五行"中属火，故在元代始建时以烽火台样式相呼应。燕墩在元代始建时，只是一座土台，位置在大都丽正门外。至明嘉靖三十二年（1553年）北京修筑外城时，才包砌以砖。

　　燕墩是一座下广上狭、平面呈正方形的墩台，台底各边长14.87米，台面各边长13.7米，台底至台面高约9米。台顶四周原有堞墙，现已损毁。墩台西北角有石门两扇，门内有石阶45级可达台顶。台顶正中是一座正方形石坛，坛上立四方石碑一座，高7.5米，面宽1.6米。碑下部为束腰须弥座，台座四周各雕花纹五层，分别为云、龙、菩提珠、菩提叶等图案，束腰部分用高浮雕技法精雕出24尊水神像。碑上部覆以四角攒尖顶，四脊各雕出一龙。碑通体汉白玉石质，南、北两面分别镌刻乾隆十八年（1753年）汉、满文字对照的《御制皇都篇》和《御制帝都篇》，皆出自乾隆手笔，是北京著名碑刻之一。

须弥座雕刻
Carvings on Shu-Mi-Tso

The Yan Pier, also known as Smoke Pier, is located at the southwest of Chongwen District, No.31 Yongdingmenwai street and the southern end of the traditional central axis, and therefore, being reputed as the landmark scenery of the southern central axis of Beijing. It was listed as a Beijing's relic under preservation in 1984.

First built in the Yuan Dynasty, the Yan Pier served as a site for emperors to pray for permanence of the imperial authority. The south belongs to "Fire" among the "Five Primary Elements", and thus buildings here were constructed into shape of watchtower in the Yuan Dynasty. At the first beginning of construction, the Yan Pier was merely an earth platform located outside Lizhengmen of Dadu. It was not until the 32nd year of the reign of Emperor Jiajing of the Ming Dynasty (1553 A.D.) when the outer city wall was under construction that the Yan Pier was wrapped with bricks.

The Yan Pier is a square platform with broad base and narrow top. The length of each side of the base is 14.87 meters, that of the top is 13.7 meters, and the height from the bottom to the top is about 9 meters. Two stone gates are opened at the northwestern corner and 45 stone steps within the gate lead straightly to the top. At the center, it stands a square stone alter, on which a 7.5-meter-high and 1.6-meter-wide stone stele is erected. At the bottom of the stele is a shu-mi-tso carved with five-tier patterns of cloud, dragon, bodhi bead and bodhi leaf etc. The girdling part is engraved with 24 sculptures of Water God by adopting the high relief technique. The upper part of the stele is covered with a pyramidal roof with four sloping ridges caved with dragons. Made up of white marble, the stele is carved with *The Chapter of Imperial Capitals* and *The Chapter of Empire Capitals* in both Chinese and Manchu in the 18th year of the reign of Emperor Qianlong (1753 A.D.) on the southern and northern sides respectively. Being the handwriting of Emperor Qianlong, the stele inscription is a well-known scenic spot of Beijing.

燕墩
Yan Pier

# 健锐营演武厅
## Ancient Qing Dynasty Fortress

健锐营演武厅位于北京市海淀区香山南路红旗村1号，俗称小团城。是清代皇帝训练及检阅八旗军的场所，亦是研究清代历史、建筑艺术、建筑规划、建筑设计最好的实物资料。2006年被公布为全国重点文物保护单位。

健锐营演武厅始建于清乾隆十三年(1748年)，是清代征服四川大小金川、稳定疆域的真实写照。健锐营及演武厅为城堡式建筑，外有护城河围绕，建筑群格局基本完整，现有面积40000平方米，是北京地区仅存的集城池、殿宇、亭台、教场为一体的武备建筑群。其主要建筑自北向南依次为小石桥、团城及演武厅和东西朝房、西城门楼、实胜寺碑亭和放马黄城(已毁)等。

团城呈椭圆形，东西约50米，南北约40米，城高11米，宽5米，周长仅190米，堪称世界上最小的城，又称看城，是大臣们观看健锐营演习、操练的地方。城墙为砖砌，墙体外侧建有雉堞，内侧为女儿墙。城内为一圈空地，青砖铺墁，有东西值房各三间，东、西城垣各设有一条马道登城。团城有城门二座，南北城门上端的玉石门额上分别雕刻着"威宣壁垒"、"志喻金汤"，均为乾隆御笔。城上南北各有一楼，南城楼面阔五间，重檐歇山顶，绿琉璃瓦屋面，四周回廊；北城楼面阔三间，重檐歇山顶，绿琉璃瓦屋面，四周回廊，内有乾隆御书汉、满、蒙、藏四种文字《实胜寺后记》碑，以表彰健锐营平定准噶尔的战功。

演武厅位于团城南侧，面阔五间，前出抱厦三间，单檐歇山顶，黄剪边绿琉璃瓦屋面，四周回廊，前有月台，乾隆皇帝曾多次在这里阅兵，厅南为占地约300亩的校场。

西城门楼俗称梯子楼，位于团城西南，虎皮石砌成，高11.2米，面阔24米，拱形门洞，南北两端各有踏道，健锐云梯营在此演练架云梯攻城攻碉。

团城南向数百米原建有庆功的实胜寺，今寺已无存，仅剩碑亭一座，重檐歇山顶，黄琉璃瓦屋面，亭内有用满、汉、蒙、藏四体文字篆刻的方形石碑，上为乾隆亲笔题书的《御制实胜寺碑记》，记述了平定大小金川的经过及实胜寺沿革。

演武厅侧面
The Side of the Drill Hall

Located at Xiangshan Nanlu in Haidian District, the Ancient Qing Dynasty Fortress, being the only one of its kind, was a military training compound for emperors of the Qing Dynasty to train and inspect their Manchu troops. It provides valuable materials for the research on history, architectural art, architectural layout and architectural design. It was listed as a national key relic under special preservation in 2006.

First built in the 13th year of the reign of Emperor Qianlong of the Qing Dynasty (1748 A.D.), the Ancient Qing Dynasty Fortress were modeled after those used by rebels in mountainous Sichuan. Manchu soldiers were selected to practice their scaling skills there. Covering an area of 40,000 square meters, this base is Beijing's only military training complex combining city, hall, pavilion and drill ground into a whole set. The principal structures, from the north to the south, are, in turn, the stone bridge, Tuancheng, the drill hall, the Shisheng Temple Stele Pavilion and so on.

Tuancheng is in the shape of an ellipse, about 50 meters from the east to the west and 40 meters from the north to the south. The circumference of the 11-meter-high and 5-meter-wide city wall is merely 190 meters, known as the minimum city in the world. Inside the city, you will find yourself in an opening, the floor of which is paved with flagstones. On the eastern and western sides, there are guardhouses with 3 bays wide respectively. There are two bridle paths leading to the top of the wall. The southern and northern sides of the city have respectively a gate with a tablet inscribed by Emperor Qianlong hung on it — the north tablet reads Zhi Yu Jin Tang (will strong as the impregnable fortress), and the south one reads Wei Xuan Bi Lei (power running over rampart). On the southern and northern gates are two gate towers respectively. The southern gate tower is 5 bays wide with a double-eaved gable-and-hip roof. The northern gate tower is 3 bays wide with a double-eaved gable-and-hip roof. The roofs of them are all covered with green glazed tiles.

Located to the south of Tuancheng, the drill hall is 5 bays wide with 3 bays in the front. The gable-and-hip roof is covered with green glazed tiles with a yellow edge. Situated to the south of the drill hall, the drill ground covers an area of about 300 Mu.

团城
Tuancheng

券门
The Arch

西城门楼
The Western Gate Tower

# 爨底下村古代建筑群
# The Ancient Village Chuandixia

爨底下村位于北京市门头沟区斋堂镇西北部，因在明代"爨里安口"险隘峡谷下方故得名。1958年简化地名用字时改"爨"为"川"。2006年被公布为全国重点文物保护单位。

爨底下村属清水河流域，四面环山，村落在峡谷北侧的缓坡上，依山而建，层层升高，整体为坐北朝南，占地约1万平方米。相传该村祖先于明朝永乐年间(1403－1424年)由山西迁移至此，建立了这座韩氏家族聚居之地。历史上，爨底下村曾为京西古驿道商品交易的客栈；解放后随着国家修建丰沙铁路和109国道，使爨底下村从商旅必经之地转为以农业为主的小山村。

现存院落74个，689间房，其中228间在抗战时期被日军烧毁，至今残存的废墟即历史的见证。一条蜿蜒的街道把村落分为上下两部分。街道用紫石、青石、灰石铺地。因依坡建筑，整个村庄的众多民宅排列高低错落有致。民宅以村北馒头山包为中心，形成南北轴线，呈扇面状向下延展，井然有序，形散而神聚，使整座村庄建筑构成和谐的整体。

Located at the northwest of Zhaitang Township in Mentougou District of Beijing, it was listed as a national key relic under special preservation in 2006.

Mountains surround Chuandixia Village and streams flow through them. The village layout merging with the slopes is quite orderly and harmonic. It, which faces south, covers an area of about 10,000 square meters. It is said that the ancestors of the village moved from Shanxi Province during the reign of Emperor Yongle of the Ming Dynasty (1403－1424 A.D.). The village used to be a flourishing inn on the post road, populated by many people. After the People's Republic of China was founded, with the development of the railway express way and advanced mountainous road the village had lost its strategic significance and not being busy any more.

The existing village comprises 74 quadrangles, having 689 buildings, 228 buildings of which were burned down by Japanese invaders. The ruins are still visible there. The village is divided into the upper and the lower parts by a wandering street paved by purple, blue and grey stones.

爨底下村鸟瞰
The Bird View of the Ancient Village Chuandixia

暴底下村一角
A Corner of the Ancient Village Chuandixia

在山坡上建房因地势不平，村民先在坡前砌石墙，逐层垫渣土夯实。根据地形而决定石墙的高度，最高有20余米。据村里老人讲石墙用石灰砌筑，在砌石墙的同时砌条石凸出墙体，为防洪水攀登而用。

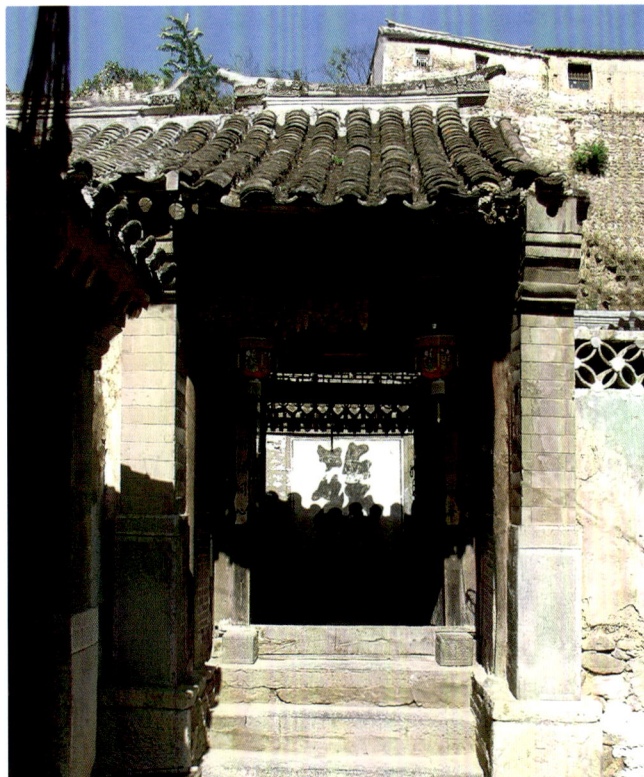

十号院门楼
The Gate of the Quadrangle at No.10

影壁
The Screen Wall

五号院正房
The Principal Room of the Quadrangle at No.5

隔扇
Partition Doors of Rooms

木雕
Woodcarvings

俯视爨底下村
Looking down the Ancient Village Chuandixia

爨底下村现存民宅以清代四合院为主体，兼有少量三合院，因地形制约，四合院往往并不规则。村民因地制宜，巧妙合理地利用有限的宅基地，将平原地区几进的大四合院，演变成为几个相对独立的各有院门的小四合院。但院与院之间留门或以0.5米至0.6米的小道相通，以便家族成员交往。在几个四合院外绕以垣墙，以保障对外的封闭性。建筑形式基本为硬山清水脊，板瓦石望板。大门楼及主体建筑前脸磨砖对缝，余部墙体为四角硬部采用细腻质坚的青砖。正房及倒座一般为三间，东西厢房各两间。台阶上均铺阶条石，正房台阶高1米多，以保障防潮，并有利于通风和采光。院内方砖铺地，各家的正房东山墙都有泰山柱支撑。爨底下村的中心是过去富户居住区。

# 附录
## Appendices

# 北京市各级别文物保护单位名录
## A List of Protected Residences of Various Levels of Beijing

### 国家级文物保护单位

| 总序号 | 保护单位名称 | 公布时间 | 市 | 县(区) | 地点 | 年代 | 类别 |
|---|---|---|---|---|---|---|---|
| 1 | 周口店遗址 | 1961.3.4 | 北京市 | 房山区 | 周口店 | 旧石器时代 | 古遗址 |
| 2 | 琉璃河遗址 | 1988.1.13 | 北京市 | 房山区 | 董家林村 | 西周 | 古遗址 |
| 3 | 圆明园遗址 | 1988.1.13 | 北京市 | 海淀区 | 清华西路28号 | 清 | 古遗址 |
| 4 | 金中都水关遗址 | 2001.6.25 | 北京市 | 丰台区 | 右安门外玉林里甲40号 | 金 | 古遗址 |
| 5 | 十三陵 | 1961.3.4 | 北京市 | 昌平区 | 昌平县境内燕山山麓 | 明 | 古墓葬 |
| 6 | 景泰陵 | 2001.6.25 | 北京市 | 海淀区 | 玉泉山北麓金山口 | 明 | 古墓葬 |
| 7 | 房山云居寺塔及石经 | 1961.3.4 | 北京市 | 房山区 | 尚乐水头村 | 隋—唐、辽、金 | 古建筑 |
| 8 | 妙应寺白塔 | 1961.3.4 | 北京市 | 西城区 | 阜成门内大街北侧 | 元 | 古建筑 |
| 9 | 真觉寺金刚宝座(五塔寺塔) | 1961.3.4 | 北京市 | 海淀区 | 西直门外长河北岸五塔寺村24号 | 明 | 古建筑 |
| 10 | 居庸关云台 | 1961.3.4 | 北京市 | 昌平区 | 居庸关关城内 | 元 | 古建筑 |
| 11 | 故宫 | 1961.3.4 | 北京市 | 东城区 | 景山前街4号 | 明—清 | 古建筑 |
| 12 | 万里长城八达岭 | 1961.3.4 | 北京市 | 延庆县 | 北京西北60公里处 | 明 | 古建筑 |
| 13 | 天坛 | 1961.3.4 | 北京市 | 崇文区 | 永定门内大街东侧 | 明 | 古建筑 |
| 14 | 北海及团城 | 1961.3.4 | 北京市 | 西城区 | 文津街1号 | 明—清 | 古建筑 |
| 15 | 智化寺 | 1961.3.4 | 北京市 | 东城区 | 禄米仓街5号 | 明 | 古建筑 |
| 16 | 国子监 | 1961.3.4 | 北京市 | 东城区 | 国子监街15号 | 清 | 古建筑 |
| 17 | 雍和宫 | 1961.3.4 | 北京市 | 东城区 | 雍和宫大街12号 | 清 | 古建筑 |
| 18 | 颐和园 | 1961.3.4 | 北京市 | 海淀区 | 北京西北郊 | 清 | 古建筑 |
| 19 | 皇史宬 | 1982.2.23 | 北京市 | 东城区 | 南池子大街136号 | 明 | 古建筑 |
| 20 | 古观象台 | 1982.2.23 | 北京市 | 东城区 | 东裱褙胡同2号 | 明—清 | 古建筑 |
| 21 | 北京城东南角楼 | 1982.2.23 | 北京市 | 东城区 | 崇文门东大街 | 明 | 古建筑 |
| 22 | 恭王府及花园 | 1982.2.23 | 北京市 | 西城区 | 前海西街17号 | 清 | 古建筑 |
| 23 | 正阳门 | 1988.1.13 | 北京市 | 东城区 | 天安门广场南侧 | 明—清 | 古建筑 |
| 24 | 太庙 | 1988.1.13 | 北京市 | 东城区 | 天安门东侧 | 明—清 | 古建筑 |
| 25 | 社稷坛 | 1988.1.13 | 北京市 | 东城区 | 天安门西侧 | 明—清 | 古建筑 |
| 26 | 北京孔庙 | 1988.1.13 | 北京市 | 东城区 | 国子监街13号 | 元—清 | 古建筑 |
| 27 | 崇礼住宅 | 1988.1.13 | 北京市 | 东城区 | 东四六条63~65号 | 清 | 古建筑 |
| 28 | 法海寺 | 1988.1.13 | 北京市 | 石景山区 | 模式口翠微山南麓 | 明 | 古建筑 |
| 29 | 牛街礼拜寺 | 1988.1.13 | 北京市 | 宣武区 | 牛街88号 | 明—清 | 古建筑 |
| 30 | 天宁寺塔 | 1988.1.13 | 北京市 | 宣武区 | 广安门外天宁寺东里2号 | 辽 | 古建筑 |
| 31 | 银山塔林 | 1988.1.13 | 北京市 | 昌平区 | 银山南麓古延寿寺遗址上 | 金—元 | 古建筑 |
| 32 | 戒台寺 | 1996.11.20 | 北京市 | 门头沟区 | 马鞍山麓 | 辽—清 | 古建筑 |
| 33 | 北京东岳庙 | 1996.11.20 | 北京市 | 朝阳区 | 朝外大街141号 | 元—清 | 古建筑 |
| 34 | 大高玄殿 | 1996.11.20 | 北京市 | 西城区 | 景山西街21、23号 | 明 | 古建筑 |
| 35 | 历代帝王庙 | 1996.11.20 | 北京市 | 西城区 | 阜成门内大街131号 | 明—清 | 古建筑 |
| 36 | 北京鼓楼、钟楼 | 1996.11.20 | 北京市 | 东城区 | 地安门外以北 | 明—清 | 古建筑 |
| 37 | 南堂 | 1996.11.20 | 北京市 | 西城区 | 前门西大街141号 | 清 | 古建筑 |
| 38 | 觉生寺 | 1996.11.20 | 北京市 | 海淀区 | 魏公村以东、北三环路北侧 | 清 | 古建筑 |
| 39 | 潭柘寺 | 2001.6.25 | 北京市 | 门头沟区 | 潭柘山 | 清 | 古建筑 |
| 40 | 可园 | 2001.6.25 | 北京市 | 东城区 | 帽儿胡同7、9、11号 | 清 | 古建筑 |
| 41 | 孚王府 | 2001.6.25 | 北京市 | 朝阳区 | 朝阳门内大街北侧 | 清 | 古建筑 |

续表：

| 总序号 | 保护单位名称 | 公布时间 | 市 | 县(区) | 地点 | 年代 | 类别 |
|---|---|---|---|---|---|---|---|
| 42 | 景山 | 2001.6.25 | 北京市 | 西城区 | 景山前街北侧 | 明—清 | 古建筑 |
| 43 | 白云观 | 2001.6.25 | 北京市 | 西城区 | 西便门外西侧 | 清 | 古建筑 |
| 44 | 万佛堂、孔水洞石刻及塔 | 2001.6.25 | 北京市 | 房山区 | 云蒙山南麓、今河北镇万佛堂村西约200米处 | 隋—唐—明 | 古建筑 |
| 45 | 法源寺 | 2001.6.25 | 北京市 | 宣武区 | 法源寺前街7号 | 清 | 古建筑 |
| 46 | 先农坛 | 2001.6.25 | 北京市 | 宣武区 | 东经路21号 | 明—清 | 古建筑 |
| 47 | 碧云寺 | 2001.6.25 | 北京市 | 海淀区 | 四季青乡寿安山东麓 | 明—清 | 古建筑 |
| 48 | 大慧寺 | 2001.6.25 | 北京市 | 海淀区 | 魏公村大慧寺路10号（钢铁研究院院内） | 明 | 古建筑 |
| 49 | 十方普觉寺 | 2001.6.25 | 北京市 | 海淀区 | 香山寿安山南麓 | 清 | 古建筑 |
| 50 | 清净化城塔 | 2001.6.25 | 北京市 | 朝阳区 | 安外西黄寺大街中部路北 | 清 | 古建筑 |
| 51 | 长城——司马台段 | 2001.6.25 | 北京市 | 密云县 | 古北口长城东约10公里，以司马明台水库为中心，向东西延伸 | | 古建筑 |
| 52 | 北京大学红楼 | 1961.3.4 | 北京市 | 东城区 | 五四大街29号 | | 近现代重要史迹及代表性建筑 |
| 53 | 卢沟桥 | 1961.3.4 | 北京市 | 丰台区 | 天安门西南十五公里的永定河上 | | 近现代重要史迹及代表性建筑 |
| 54 | 天安门 | 1961.3.4 | 北京市 | 东城区 | 天安门广场 | | 近现代重要史迹及代表性建筑 |
| 55 | 人民英雄纪念碑 | 1961.3.4 | 北京市 | 东城区 | 天安门广场 | 1958年 | 近现代重要史迹及代表性建筑 |
| 56 | 北京宋庆龄故居 | 1982.2.23 | 北京市 | 西城区 | 后海北沿46号 | 1963—1981年 | 近现代重要史迹及代表性建筑 |
| 57 | 郭沫若故居 | 1988.1.13 | 北京市 | 西城区 | 前海西街18号 | 1963—1978年 | 近现代重要史迹及代表性建筑 |
| 58 | 东交民巷使馆建筑群 | 2001.6.25 | 北京市 | 东城区 | 近代 | | 近现代重要史迹及 代表性建筑 |
| 59 | 未名湖燕园建筑 | 2001.6.25 | 北京市 | 海淀区 | 北京大学校园内 | 近代 | 近现代重要史迹及代表性建筑 |
| 60 | 清华大学早期建筑 | 2001.6.25 | 北京市 | 海淀区 | 清华大学校园内 | 近代 | 近现代重要史迹及代表性建筑 |
| 61 | 十字寺遗址 | 2006.5.25 | 北京市 | 房山区 | 周口店镇车厂村北三盆山 | 元 | 古遗址 |
| 62 | 元大都城墙遗址 | 2006.5.25 | 北京市 | 朝阳区 | 海淀区 | 元 | 古遗址 |
| 63 | 金陵 | 2006.5.25 | 北京市 | 房山区 | 周口店镇车厂村 | 金 | 古墓葬 |
| 64 | 利玛窦和外国传教士墓地 | 2006.5.25 | 北京市 | 西城区 | 车公庄大街6号 | 明—清 | 古墓葬 |
| 65 | 袁崇焕墓和祠 | 2006.5.25 | 北京市 | 崇文区 | 东花市斜街52号、龙潭路8号 | 明—民国 | 古墓葬 |
| 66 | 承恩寺 | 2006.5.25 | 北京市 | 石景山区 | 模式口街 | 明—清 | 古建筑 |
| 67 | 大觉寺 | 2006.5.25 | 北京市 | 海淀区 | 北安河乡 | 明—清 | 古建筑 |
| 68 | 德胜门箭楼 | 2006.5.25 | 北京市 | 西城区 | 德胜门立交桥北侧 | 明—清 | 古建筑 |
| 69 | 地坛 | 2006.5.25 | 北京市 | 东城区 | 安定门外大街东侧 | 明—清 | 古建筑 |
| 70 | 日坛 | 2006.5.25 | 北京市 | 朝阳区 | 朝阳门外日坛公园 | 明—清 | 古建筑 |
| 71 | 月坛 | 2006.5.25 | 北京市 | 西城区 | 月坛北街南侧 | 明—清 | 古建筑 |
| 72 | 中南海 | 2006.5.25 | 北京市 | 西城区 | 西长安街北侧 | 明—清 | 古建筑 |
| 73 | 安徽会馆 | 2006.5.25 | 北京市 | 宣武区 | 后孙公园17、19、23、25、27号 | 清 | 古建筑 |
| 74 | 柏林寺 | 2006.5.25 | 北京市 | 东城区 | 戏楼胡同1号 | 清 | 古建筑 |
| 75 | 报国寺 | 2006.5.25 | 北京市 | 宣武区 | 报国寺前街1号 | 清 | 古建筑 |
| 76 | 醇亲王府 | 2006.5.25 | 北京市 | 西城区 | 后海北沿44号、鼓楼西大街154、156号 | 清 | 古建筑 |
| 77 | 爨底下村古建筑群 | 2006.5.25 | 北京市 | 门头沟区 | 斋堂镇爨底下村 | 清 | 古建筑 |
| 78 | 关岳庙 | 2006.5.25 | 北京市 | 西城区 | 鼓楼西大街149号 | 清 | 古建筑 |
| 79 | 广济寺 | 2006.5.25 | 北京市 | 西城区 | 阜成门内大街25号 | 清 | 古建筑 |
| 80 | 健锐营演武厅 | 2006.5.25 | 北京市 | 海淀区 | 香山南路红旗村1号 | 清 | 古建筑 |
| 81 | 静明园 | 2006.5.25 | 北京市 | 海淀区 | 玉泉山 | 清 | 古建筑 |
| 82 | 万寿寺 | 2006.5.25 | 北京市 | 海淀区 | 苏州街长河北岸 | 清 | 古建筑 |
| 83 | 国立蒙藏学校旧址 | 2006.5.25 | 北京市 | 西城区 | 小石虎胡同33号 | 清 | 近现代重要史迹及代表性建筑 |
| 84 | 京师大学堂分科大学旧址 | 2006.5.25 | 北京市 | 东城区 | 安德里北街21号 | 清 | 近现代重要史迹及代表性建筑 |

续表：

| 总序号 | 保护单位名称 | 公布时间 | 市 | 县(区) | 地点 | 年代 | 类别 |
|---|---|---|---|---|---|---|---|
| 85 | 清陆军部和海军部旧址 | 2006.5.25 | 北京市 | 东城区 | 张自忠路3号 | 清 | 近现代重要史迹及代表性建筑 |
| 86 | 清农事试验场旧址 | 2006.5.25 | 北京市 | 西城区 | 西直门外大街137号北京动物园内 | 清 | 近现代重要史迹及代表性建筑 |
| 87 | 西什库教堂 | 2006.5.25 | 北京市 | 西城区 | 西什库大街33号 | 清 | 近现代重要史迹及代表性建筑 |
| 88 | 亚斯立堂 | 2006.5.25 | 北京市 | 东城区 | 后沟胡同丁2号 | 清 | 近现代重要史迹及代表性建筑 |
| 89 | 北京国会旧址 | 2006.5.25 | 北京市 | 西城区 | 宣武门西大街57号、佟麟阁路62号 | 清—民国 | 近现代重要史迹及代表性建筑 |
| 90 | 大栅栏商业建筑 | 2006.5.25 | 北京市 | 宣武区 | 大栅栏街1、5号、珠宝市街5号、前门廊房头条17号 | 清—民国 | 近现代重要史迹及代表性建筑 |
| 91 | 国民政府财政部印刷局旧址 | 2006.5.25 | 北京市 | 宣武区 | 白纸坊路23号 | 清—民国 | 近现代重要史迹及代表性建筑 |
| 92 | 京师女子师范学堂旧址 | 2006.5.25 | 北京市 | 西城区 | 新文化街45号 | 清—民国 | 近现代重要史迹及代表性建筑 |
| 93 | 协和医学院旧址 | 2006.5.25 | 北京市 | 东城区 | 帅府园胡同1号 | 清—民国 | 近现代重要史迹及代表性建筑 |
| 94 | 北平图书馆旧址 | 2006.5.25 | 北京市 | 西城区 | 文津街7号 | 民国 | 近现代重要史迹及代表性建筑 |
| 95 | 辛亥滦州起义纪念园 | 2006.5.25 | 北京市 | 海淀区 | 温泉村显龙山 | 民国 | 近现代重要史迹及代表性建筑 |
| 96 | 孙中山行馆 | 2006.5.25 | 北京市 | 东城区 | 张自忠路23号 | 1924—1925年 | 近现代重要史迹及代表性建筑 |
| 97 | 北京鲁迅旧居 | 2006.5.25 | 北京市 | 西城区 | 阜成门内北京鲁迅博物馆内 | 1924—1926年 | 近现代重要史迹及代表性建筑 |
| 98 | 京杭大运河 | 2006.5.25 | 北京市 | | | 春秋—清 | 古建筑 |

# 市级文物保护单位

## 宣武区

| 总序号 | 保护单位名称 | 公布时间 | 市 | 县(区) | 地点 | 年代 | 类别 |
|---|---|---|---|---|---|---|---|
| 1 | 金中都太液池遗址 | 1984.5.24 | 北京市 | 宣武区 | 广安门外南街77号 | 金 | 古遗址 |
| 2 | 陶然亭慈悲庵 | 1979.8.21 | 北京市 | 宣武区 | 陶然亭公园内 | 明—清 | 古建筑 |
| 3 | 中山会馆 | 1984.5.24 | 北京市 | 宣武区 | 珠巢街5号 | 明 | 古建筑 |
| 4 | 杨椒山祠(松筠庵) | 1984.5.24 | 北京市 | 宣武区 | 宣武门外达智桥12号、校场口三条2号 | 明—清 | 古建筑 |
| 5 | 湖广会馆 | 1984.5.24 | 北京市 | 宣武区 | 虎坊路3、5号 | 清 | 古建筑 |
| 6 | 云绘楼清音阁 | 1984.5.24 | 北京市 | 宣武区 | 陶然亭公园内 | 清 | 古建筑 |
| 7 | 朱彝尊故居(顺德会馆) | 1984.5.24 | 北京市 | 宣武区 | 海柏胡同16号 | 清 | 古建筑 |
| 8 | 长椿寺 | 2001.7.12 | 北京市 | 宣武区 | 长椿街9、11号 | 明 | 古建筑 |
| 9 | 正乙祠 | 2001.7.12 | 北京市 | 宣武区 | 西河沿街220号 | 清 | 古建筑 |
| 10 | 三圣庵 | 2001.7.12 | 北京市 | 宣武区 | 陶然亭北里黑窑厂14号 | 清末 | 古建筑 |
| 11 | 纪晓岚故居 | 2003.12.11 | 北京市 | 宣武区 | 珠市口西大街241号 | 清 | 古建筑 |
| 12 | 湖南会馆 | 1984.5.24 | 北京市 | 宣武区 | 烂缦胡同101、103号 | 清 | 近现代重要史迹及代表性建筑 |
| 13 | 康有为故居 | 1984.5.24 | 北京市 | 宣武区 | 米市胡同43号 | 清 | 近现代重要史迹及代表性建筑 |
| 14 | 京报馆 | 1984.5.24 | 北京市 | 宣武区 | 魏染胡同30、32号 | 民国 | 近现代重要史迹及代表性建筑 |
| 15 | 盐业银行旧址 | 1995.10.20 | 北京市 | 宣武区 | 前门西河沿7号 | 1931年 | 近现代重要史迹及代表性建筑 |
| 16 | 交通银行旧址 | 1995.10.20 | 北京市 | 宣武区 | 前门西河沿9号 | 1937年 | 近现代重要史迹及代表性建筑 |
| 17 | 粮食店街第十旅馆 | 2001.7.12 | 北京市 | 宣武区 | 粮食店街73号 | 民国 | 近现代重要史迹及代表性建筑 |
| 18 | 京华印书局 | 2003.12.11 | 北京市 | 宣武区 | 虎坊桥北口 | 1920年 | 近现代重要史迹及代表性建筑 |
| 19 | 德寿堂药店 | 2003.12.11 | 北京市 | 宣武区 | 珠市口西大街175号 | 1934年 | 近现代重要史迹及代表性建筑 |

## 崇文区

| 总序号 | 保护单位名称 | 公布时间 | 市 | 县(区) | 地点 | 年代 | 类别 |
|---|---|---|---|---|---|---|---|
| 1 | 福建汀州会馆北馆 | 1984.5.24 | 北京市 | 崇文区 | 长巷二条46、48号 | 明 | 古建筑 |

续表：

| 总序号 | 保护单位名称 | 公布时间 | 市 | 县(区) | 地点 | 年代 | 类别 |
|---|---|---|---|---|---|---|---|
| 2 | 隆安寺 | 1984.5.24 | 北京市 | 崇文区 | 白桥南里1、3号 | 明—清 | 古建筑 |
| 3 | 金台书院 | 1984.5.24 | 北京市 | 崇文区 | 东晓市街203号 | 清 | 古建筑 |
| 4 | 阳平会馆戏楼 | 1984.5.24 | 北京市 | 崇文区 | 小江胡同34、36号 | 清 | 古建筑 |
| 5 | 花市火神庙 | 2003.12.11 | 北京市 | 崇文区 | 西花市大街113号 | 清 | 古建筑 |
| 6 | 燕墩 | 1984.5.24 | 北京市 | 崇文区 | 永定门外大街31号 | 清 | 石窟寺及石刻 |
| 7 | 正阳桥疏渠记方碑 | 1984.5.24 | 北京市 | 崇文区 | 天桥红庙街78号 | 清 | 石窟寺及石刻 |
| 8 | 崇文区新革路20号四合院 | 1984.5.24 | 北京市 | 崇文区 | 新革路20号 | 民国 | 近现代重要史迹及代表性建筑 |
| 9 | 京奉铁路正阳门东车站旧址 | 2001.7.12 | 北京市 | 崇文区 | 前门大街东侧 | 1903年 | 近现代重要史迹及代表性建筑 |

## 东城区

| 总序号 | 保护单位名称 | 公布时间 | 市 | 县(区) | 地点 | 年代 | 类别 |
|---|---|---|---|---|---|---|---|
| 1 | 文天祥祠 | 1979.8.21 | 北京市 | 东城区 | 府学胡同63号 | 明—清 | 古建筑 |
| 2 | 国子监街 | 1984.5.24 | 北京市 | 东城区 | 国子监街 | 元—清 | 古建筑 |
| 3 | 东四清真寺 | 1984.5.24 | 北京市 | 东城区 | 东四南大街13号 | 明 | 古建筑 |
| 4 | 北新仓 | 1984.5.24 | 北京市 | 东城区 | 北新仓胡同甲16号 | 明—清 | 古建筑 |
| 5 | 禄米仓 | 1984.5.24 | 北京市 | 东城区 | 禄米仓胡同71、73号 | 明—清 | 古建筑 |
| 6 | 南新仓 | 1984.5.24 | 北京市 | 东城区 | 朝阳门东四十条22号 | 明—清 | 古建筑 |
| 7 | 顺天府学 | 1984.5.24 | 北京市 | 东城区 | 府学胡同65号 | 明—清 | 古建筑 |
| 8 | 东城区方家胡同13、15号四合院 | 1984.5.24 | 北京市 | 东城区 | 方家胡同13、15号 | 清 | 古建筑 |
| 9 | 东城区府学胡同36号（包括交道口南大街136号）四合院 | 1984.5.24 | 北京市 | 东城区 | 府学胡同36号、交道口南大街136号 | 清 | 古建筑 |
| 10 | 东城区国祥胡同2号四合院 | 1984.5.24 | 北京市 | 东城区 | 国祥胡同甲2号 | 清 | 古建筑 |
| 11 | 东城区礼士胡同 | 1984.5.24 | 北京市 | 东城区 | 礼士胡同129号 | 清 | 古建筑 |
| 12 | 东城区内务部街11号四合院 | 1984.5.24 | 北京市 | 东城区 | 内务部街11号 | 清 | 古建筑 |
| 13 | 东城区圆恩寺后街7号四合院 | 1984.5.24 | 北京市 | 东城区 | 圆恩寺后街7、9号 | 清 | 古建筑 |
| 14 | 和敬公主府 | 1984.5.24 | 北京市 | 东城区 | 张自忠路7号 | 清 | 古建筑 |
| 15 | 旧宅院 | 1984.5.24 | 北京市 | 东城区 | 帽儿胡同35、37号 | 清 | 古建筑 |
| 16 | 凝和庙 | 1984.5.24 | 北京市 | 东城区 | 北池子46号 | 清 | 古建筑 |
| 17 | 普度寺大殿 | 1984.5.24 | 北京市 | 东城区 | 普庆前巷 | 清 | 古建筑 |
| 18 | 嵩祝寺及智珠寺 | 1984.5.24 | 北京市 | 东城区 | 嵩祝院北巷 | 清 | 古建筑 |
| 19 | 宣仁庙 | 1984.5.24 | 北京市 | 东城区 | 北池子2号 | 清 | 古建筑 |
| 20 | 于谦祠 | 1984.5.24 | 北京市 | 东城区 | 裱背胡同23号 | 清 | 古建筑 |
| 21 | 大慈延福宫建筑遗存 | 1990.2.23 | 北京市 | 东城区 | 朝阳门内大街203号 | 明 | 古建筑 |
| 22 | 东城区西堂子胡同25~37号四合院 | 1990.2.23 | 北京市 | 东城区 | 西堂子胡同25~37号 | 清 | 古建筑 |
| 23 | 东堂 | 1990.2.23 | 北京市 | 东城区 | 王府井大街74号 | 清 | 古建筑 |
| 24 | 东城区帽儿胡同5号四合院 | 2001.7.12 | 北京市 | 东城区 | 帽儿胡同5号 | 清 | 古建筑 |
| 25 | 东城区前鼓楼苑胡同7、9号四合院 | 2001.7.12 | 北京市 | 东城区 | 前鼓楼苑胡同7、9号 | 清 | 古建筑 |
| 26 | 宁郡王府 | 2001.7.12 | 北京市 | 东城区 | 北极阁三条 | 清 | 古建筑 |
| 27 | 东城区美术馆东街25号四合院 | 2001.7.12 | 北京市 | 东城区 | 美术馆东街25号 | 清末 | 古建筑 |

续表：

| 总序号 | 保护单位名称 | 公布时间 | 市 | 县(区) | 地点 | 年代 | 类别 |
|---|---|---|---|---|---|---|---|
| 28 | 东城区鼓楼东大街255号四合院 | 2001.7.12 | 北京市 | 东城区 | 鼓楼东大街255号 | 民国 | 古建筑 |
| 29 | 东城区黑芝麻胡同13号四合院 | 2003.12.11 | 北京市 | 东城区 | 黑芝麻胡同13号 | 清 | 古建筑 |
| 30 | 东城区前永康胡同7号四合院 | 2003.12.11 | 北京市 | 东城区 | 前永康胡同7号 | 清 | 古建筑 |
| 31 | 东城区沙井胡同15号四合院 | 2003.12.11 | 北京市 | 东城区 | 沙井胡同15号 | 清 | 古建筑 |
| 32 | 恒亲王府 | 2003.12.11 | 北京市 | 东城区 | 朝阳门内大街 | 清 | 古建筑 |
| 33 | 绮园花园 | 2003.12.11 | 北京市 | 东城区 | 秦老胡同35号 | 清 | 古建筑 |
| 34 | 僧王府 | 2003.12.11 | 北京市 | 东城区 | 炒豆胡同77号、板厂胡同30~34号 | 清 | 古建筑 |
| 35 | 总理各国事务衙门 | 2003.12.11 | 北京市 | 东城区 | 东堂子胡同49号 | 清 | 古建筑 |
| 36 | 毛主席故居 | 1979.8.21 | 北京市 | 东城区 | 景山东街三眼井吉安所左巷8号 | 民国 | 近现代重要史迹及代表性建筑 |
| 37 | 毛主席纪念堂 | 1979.8.21 | 北京市 | 东城区 | 天安门广场 | 1977年 | 近现代重要史迹及代表性建筑 |
| 38 | 老舍故居 | 1984.5.24 | 北京市 | 东城区 | 丰富胡同19号 | 民国 | 近现代重要史迹及代表性建筑 |
| 39 | 茅盾故居 | 1984.5.24 | 北京市 | 东城区 | 后圆恩寺胡同13号 | 民国 | 近现代重要史迹及代表性建筑 |
| 40 | 原中法大学 | 1984.5.24 | 北京市 | 东城区 | 东黄城根北街甲20号 | 民国 | 近现代重要史迹及代表性建筑 |
| 41 | 中华圣经会旧址 | 1984.5.24 | 北京市 | 东城区 | 东单北大街21号 | 民国 | 近现代重要史迹及代表性建筑 |
| 42 | 京师大学堂建筑遗存 | 1990.2.23 | 北京市 | 东城区 | 沙滩后街55、59号 | 清一民国 | 近现代重要史迹及代表性建筑 |
| 43 | 北京大学地质馆旧址 | 1990.2.23 | 北京市 | 东城区 | 沙滩北街15号 | 1931年 | 近现代重要史迹及代表性建筑 |
| 44 | 北京饭店初期建筑 | 1990.2.23 | 北京市 | 东城区 | 东长安街 | 1931年 | 近现代重要史迹及代表性建筑 |
| 45 | 美国使馆旧址 | 1995.10.20 | 北京市 | 东城区 | 前门东大街23号 | 1903年 | 近现代重要史迹及代表性建筑 |
| 46 | 圣米厄尔教堂 | 1995.10.20 | 北京市 | 东城区 | 东交民巷甲13号 | 1904年 | 近现代重要史迹及代表性建筑 |
| 47 | 荷兰使馆旧址 | 1995.10.20 | 北京市 | 东城区 | 前门东大街11号 | 1909年 | 近现代重要史迹及代表性建筑 |
| 48 | 法国邮政局旧址 | 1995.10.20 | 北京市 | 东城区东 | 东交民巷19号 | 1910年 | 近现代重要史迹及代表性建筑 |
| 49 | 军调部1946年中共代表团驻地 | 1995.10.20 | 北京市 | 东城区 | 东安门大街1号 | 1946年 | 近现代重要史迹及代表性建筑 |
| 50 | 孑民堂 | 1995.10.20 | 北京市 | 东城区 | 沙滩北街甲83号 | 1947年 | 近现代重要史迹及代表性建筑 |
| 51 | 陈独秀旧居箭杆胡同20号 | 2001.7.12 | 北京市 | 东城区 | 北池子大街 | 民国 | 近现代重要史迹及代表性建筑 |
| 52 | 东城区东棉花胡同15号院及拱门砖雕 | 2001.7.12 | 北京市 | 东城区 | 东棉花胡同15号 | 民国 | 近现代重要史迹及代表性建筑 |
| 53 | 协和医院住宅群北极阁三条26号 | 2003.12.11 | 北京市 | 东城区 | 外交部街59号、 | 近代 | 近现代重要史迹及代表性建筑 |
| 54 | 原麦加利银行 | 2003.12.11 | 北京市 | 东城区 | 东交民巷甲39号 | 清末 | 近现代重要史迹及代表性建筑 |
| 55 | 北京大学女生宿舍 | 2003.12.11 | 北京市 | 东城区 | 沙滩北街乙2号 | 1935年 | 近现代重要史迹及代表性建筑 |

## 西城区

| 总序号 | 保护单位名称 | 公布时间 | 市 | 县(区) | 地点 | 年代 | 类别 |
|---|---|---|---|---|---|---|---|
| 1 | 百万庄路8号墓园石刻 | 2001.7.12 | 北京市 | 西城区 | 百万庄路8号 | 民国 | 古墓葬 |
| 2 | 万宁桥(后门桥) | 1984.5.24 | 北京市 | 西城区 | 地安门外大街 | 元 | 古建 |
| 3 | 都城隍庙后殿(寝祠) | 1984.5.24 | 北京市 | 西城区 | 成方街33号 | 元一清 | 古建筑 |
| 4 | 广化寺 | 1984.5.24 | 北京市 | 西城区 | 后海鸦儿胡同31号 | 元一清 | 古建筑 |
| 5 | 护国寺金刚殿 | 1984.5.24 | 北京市 | 西城区 | 护国寺西巷 | 元一清 | 古建筑 |
| 6 | 火德真君庙(火神庙) | 1984.5.24 | 北京市 | 西城区 | 地安门外大街77号 | 明 | 古建筑 |

续表：

| 总序号 | 保护单位名称 | 公布时间 | 市 | 县(区) | 地点 | 年代 | 类别 |
|---|---|---|---|---|---|---|---|
| 7 | 福佑寺 | 1984.5.24 | 北京市 | 西城区 | 北长街20号 | 清 | 古建筑 |
| 8 | 克勤郡王府 | 1984.5.24 | 北京市 | 西城区 | 新文化街53号 | 清 | 古建筑 |
| 9 | 礼王府 | 1984.5.24 | 北京市 | 西城区 | 西黄城根7、9号、颁赏甲19号 | 清 | 古建筑 |
| 10 | 吕祖阁 | 1984.5.24 | 北京市 | 西城区 | 明光胡同6号、新壁胡同41号 | 清 | 古建筑 |
| 11 | 庆王府 | 1984.5.24 | 北京市 | 西城区 | 定阜大街3号、德内大街254号 | 清 | 古建筑 |
| 12 | 升平署戏楼 | 1984.5.24 | 北京市 | 西城区 | 西长安街1号 | 清 | 古建筑 |
| 13 | 昭显庙 | 1984.5.24 | 北京市 | 西城区 | 北长街71号 | 清 | 古建筑 |
| 14 | 郑王府 | 1984.5.24 | 北京市 | 西城区 | 大木仓胡同35号 | 清 | 古建筑 |
| 15 | 万松老人塔 | 1995.10.20 | 北京市 | 西城区 | 西四南大街43号旁门 | 元 | 古建筑 |
| 16 | 涛贝勒府 | 1995.10.20 | 北京市 | 西城区 | 柳荫街25、27、乙27号 | 清 | 古建筑 |
| 17 | 贤良祠 | 2001.7.12 | 北京市 | 西城区 | 地安门西大街103号 | 清 | 古建筑 |
| 18 | 旧式铺面房 | 2001.7.12 | 北京市 | 西城区 | 地安门外大街50、52号 | 清末 | 古建筑 |
| 19 | 拈花寺 | 2003.12.11 | 北京市 | 西城区 | 大石桥胡同61号 | 明 | 古建筑 |
| 20 | 恭俭冰窖 | 2003.12.11 | 北京市 | 西城区 | 恭俭五巷5号 | 清 | 古建筑 |
| 21 | 会贤堂 | 2003.12.11 | 北京市 | 西城区 | 前海北沿18号 | 清 | 古建筑 |
| 22 | 地安门 | 2003.12.11 | 北京市 | 西城区 | 地安门西大街153号 | 清 | 古建筑 |
| 23 | 雪池冰窖 | 2003.12.11 | 北京市 | 西城区 | 雪池胡同10号 | 清 | 古建筑 |
| 24 | 西城区阜成门内大街93号四合院 | 2003.12.11 | 北京市 | 西城区 | 阜成门内大街93号 | 民国 | 古建筑 |
| 25 | 李大钊故居 | 1979.8.21 | 北京市 | 西城区 | 文华胡同24号 | 民国 | 近现代重要史迹及代表性建筑 |
| 26 | 天主教圣母会法址文学校旧 | 1984.5.24 | 北京市 | 西城区 | 前门西大街137号 | 清 | 近现代重要史迹及代表性建筑 |
| 27 | 西城区西交民巷87号、北新华街112号四合院 | 1984.5.24 | 北京市 | 西城区 | 西交民巷87号、北新华街112号 | 清 | 近现代重要史迹及代表性建筑 |
| 28 | 程砚秋故居 | 1984.5.24 | 北京市 | 西城区 | 西四北三条39号 | 民国 | 近现代重要史迹及代表性建筑 |
| 29 | 梅兰芳故居 | 1984.5.24 | 北京市 | 西城区 | 护国寺街9号 | 民国 | 近现代重要史迹及代表性建筑 |
| 30 | 齐白石故居 | 1984.5.24 | 北京市 | 西城区 | 跨车胡同13号 | 民国 | 近现代重要史迹及代表性建筑 |
| 31 | 西城区前公用胡同15号四合院 | 1984.5.24 | 北京市 | 西城区 | 前公用胡同15号 | 民国 | 近现代重要史迹及代表性建筑 |
| 32 | 西城区西四北大街6条23号四合院 | 1984.5.24 | 北京市 | 西城区 | 西四北六条23号 | 民国 | 近现代重要史迹及代表性建筑 |
| 33 | 西城区西四北3条19号四合院 | 1984.5.24 | 北京市 | 西城区 | 西四北三条19号 | 民国 | 近现代重要史迹及代表性建筑 |
| 34 | 西城区西四北三条11号四合院 | 1984.5.24 | 北京市 | 西城区 | 西四北三条11号 | 民国 | 近现代重要史迹及代表性建筑 |
| 35 | 原辅仁大学 | 1984.5.24 | 北京市 | 西城区 | 定阜大街1号 | 民国 | 近现代重要史迹及代表性建筑 |
| 36 | 盛新中学与佑贞女中旧址 | 1990.2.23 | 北京市 | 西城区 | 教场胡同2、4号 | 1928年 | 近现代重要史迹及代表性建筑 |
| 37 | 西城区富国街3号四合院 | 1995.10.20 | 北京市 | 西城区 | 富国街3号 | 清 | 近现代重要史迹及代表性建筑 |
| 38 | 平绥西直门车站旧址 | 1995.10.20 | 北京市 | 西城区 | 北滨河路1号 | 1909年 | 近现代重要史迹及代表性建筑 |
| 39 | 北京水准原点旧址 | 1995.10.20 | 北京市 | 西城区 | 西安门大街1号 | 1915年 | 近现代重要史迹及代表性建筑 |
| 40 | 中国农工银行旧址 | 1995.10.20 | 北京市 | 西城区 | 西交民巷50号 | 1922年 | 近现代重要史迹及代表性建筑 |
| 41 | 大陆银行旧址 | 1995.10.20 | 北京市 | 西城区 | 西交民巷17号 | 1924年 | 近现代重要史迹及代表性建筑 |
| 42 | 中央银行旧址 | 1995.10.20 | 北京市 | 西城区 | 西交民巷17号 | 20世纪30年代 | 近现代重要史迹及代表性建筑 |
| 43 | 保商银行旧址 | 1995.10.20 | 北京市 | 西城区 | 西交民巷17号 | 20世纪30年代 | 近现代重要史迹及代表性建筑 |
| 44 | 中华圣公会教堂 | 2001.7.12 | 北京市 | 西城区 | 佟麟阁路85号 | 1907年 | 近现代重要史迹及代表性建筑 |

## 朝阳区

| 总序号 | 保护单位名称 | 公布时间 | 市 | 县(区) | 地点 | 年代 | 类别 |
|---|---|---|---|---|---|---|---|
| 1 | 永通桥及石道碑 | 1984.5.24 | 北京市 | 朝阳区 | 管庄乡八里桥村 | 明—清 | 古建筑 |
| 2 | 顺承郡王府 | 1984.5.24 | 北京市 | 朝阳区 | 朝阳公园东侧 | 清 | 古建筑 |
| 3 | 十方诸佛宝塔 | 1990.2.23 | 北京市 | 朝阳区 | 王四营乡马房寺村 | 明 | 古建筑 |
| 4 | 北顶娘娘庙 | 2003.12.11 | 北京市 | 朝阳区 | 奥运村游泳馆南 | 明 | 古建筑 |
| 5 | 四九一电台旧址 | 2001.7.12 | 北京市 | 朝阳区 | 双桥1号 | 1918年 | 近现代重要史迹及代表性建筑 |

## 海淀区

| 总序号 | 保护单位名称 | 公布时间 | 市 | 县(区) | 地点 | 年代 | 类别 |
|---|---|---|---|---|---|---|---|
| 1 | 清河汉城遗址 | 2001.7.12 | 北京市 | 海淀区 | 东升乡朱房村 | 汉 | 古遗址 |
| 2 | 醇亲王墓 | 1984.5.24 | 北京市 | 海淀区 | 苏家坨镇七王坟路17号 | 清 | 古墓葬 |
| 3 | 孚郡王墓 | 1984.5.24 | 北京市 | 海淀区 | 苏家坨镇草场村西南 | 清 | 古墓葬 |
| 4 | 慈寿寺塔 | 1957.10.28 | 北京市 | 海淀区 | 阜外八里庄北里3号 | 明 | 古建筑 |
| 5 | 静宜园(香山) | 1984.5.24 | 北京市 | 海淀区 | 香山公园 | 辽—清 | 古建筑 |
| 6 | 广济桥(清河大桥) | 1984.5.24 | 北京市 | 海淀区 | 清河镇 | 明 | 古建筑 |
| 7 | 黑龙潭及龙王庙 | 1984.5.24 | 北京市 | 海淀区 | 白家 | 明—清 | 古建筑 |
| 8 | 钓鱼台与养源斋 | 1984.5.24 | 北京市 | 海淀区 | 阜成路2号 | 清 | 古建筑 |
| 9 | 乐家花园 | 1984.5.24 | 北京市 | 海淀区 | 海淀镇八一中学 | 清 | 古建筑 |
| 10 | 旭华之阁及松堂 | 1984.5.24 | 北京市 | 海淀区 | 香山南路红旗村8号 | 清 | 古建筑 |
| 11 | 定慧寺 | 1990.2.23 | 北京市 | 海淀区 | 阜成路66号 | 明 | 古建筑 |
| 12 | 摩诃庵 | 1995.10.20 | 北京市 | 海淀区 | 八里庄街37号 | 明 | 古建筑 |
| 13 | 广仁宫(西顶) | 2001.7.12 | 北京市 | 海淀区 | 四季青镇蓝靛厂 | 明—清 | 古建筑 |
| 14 | 上庄东岳庙 | 2003.12.11 | 北京市 | 海淀区 | 上庄镇永泰庄 | 清 | 古建筑 |
| 15 | 魏太和造像 | 1957.10.28 | 北京市 | 海淀区 | 苏家坨镇东耳营村 | 北魏 | 石窟寺及石刻 |
| 16 | 双清别墅 | 1979.8.21 | 北京市 | 海淀区 | 香山公园内 | 民国 | 近现代重要史迹及代表性建筑 |
| 17 | "三一八"烈士纪念碑 | 1984.5.24 | 北京市 | 海淀区 | 圆明园"九州清宴"遗址内 | 民国 | 近现代重要史迹及代表性建筑 |
| 18 | 达园 | 1984.5.24 | 北京市 | 海淀区 | 福缘门甲1号 | 民国 | 近现代重要史迹及代表性建筑 |
| 19 | 李大钊烈士陵园 | 1984.5.24 | 北京市 | 海淀区 | 万安公墓内 | 1983年 | 近现代重要史迹及代表性建筑 |
| 20 | 梁启超墓 | 2001.7.12 | 北京市 | 海淀区 | 北京植物园 | 1931年 | 近现代重要史迹及代表性建筑 |
| 21 | 孙岳墓 | 2003.12.11 | 北京市 | 海淀区 | 温泉北京胸科医院 | 近代 | 近现代重要史迹及代表性建筑 |

## 石景山区

| 总序号 | 保护单位名称 | 公布时间 | 市 | 县(区) | 地点 | 年代 | 类别 |
|---|---|---|---|---|---|---|---|
| 1 | 老山汉墓 | 2001.7.12 | 北京市 | 石景山区 | 老山东南麓 | 西汉 | 古墓葬 |
| 2 | 田义墓 | 2001.7.12 | 北京市 | 石景山区 | 模式口大街北侧 | 明 | 古墓葬 |
| 3 | 西山八大处(长安寺、灵光寺、三山庵、大悲庵、龙王堂、香界寺、宝珠洞、证果寺) | 1957.10.28 | 北京市 | 石景山区 | 八大处 | 明—清 | 古建筑 |
| 4 | 慈善寺 | 1995.10.20 | 北京市 | 石景山区 | 天台山 | 清 | 古建筑 |
| 5 | 北京八宝山革命公墓 | 1984.5.24 | 北京市 | 石景山区 | 八宝山地铁东侧 | 1950年 | 近现代重要史迹及代表性建筑 |
| 6 | 冰川擦痕 | 1957.10.28 | 北京市 | 石景山区 | 模式口 | 第四纪 | 其他 |

## 丰台区

| 总序号 | 保护单位名称 | 公布时间 | 市 | 县(区) | 地点 | 年代 | 类别 |
|---|---|---|---|---|---|---|---|
| 1 | 金中都城遗迹 | 1984.5.24 | 北京市 | 丰台区 | 卢沟桥三路居 凤凰嘴村、高楼村、万泉寺村 | 金 | 古遗址 |
| 2 | 莲花池 | 1984.5.24 | 北京市 | 丰台区 | 广外莲花池路48号 | 金 | 古遗址 |

续表：

| 总序号 | 保护单位名称 | 公布时间 | 市 | 县(区) | 地点 | 年代 | 类别 |
|---|---|---|---|---|---|---|---|
| 3 | 大葆台西汉墓遗址 | 1995.10.20 | 北京市 | 丰台区 | 丰葆路世界公园南680米处 | 西汉 | 古墓葬 |
| 4 | 镇岗塔 | 1957.10.28 | 北京市 | 丰台区 | 长辛店镇张家坟村 | 金 | 古建筑 |
| 5 | 南岗洼桥 | 2001.7.12 | 北京市 | 丰台区 | 王佐镇南岗洼村南 | 明末清初 | 古建筑 |
| 6 | 丰台娘娘庙 | 2003.12.11 | 北京市 | 丰台区 | 长辛店镇大灰厂村 | 明 | 古建筑 |
| 7 | 丰台药王庙 | 2003.12.11 | 北京市 | 丰台区 | 花乡看丹村 | 清 | 古建筑 |
| 8 | 长辛店"二七"革命遗址 | 1979.8.21 | 北京市 | 丰台区 | 长辛店镇祠堂口1号、长辛店大街174号 | 民国 | 近现代重要史迹及代表性建筑 |
| 9 | 长辛店留法勤工俭学旧址 | 1984.5.24 | 北京市 | 丰台区 | 长辛店德善里18号铁路中学院内 | 民国 | 近现代重要史迹及代表性建筑 |
| 10 | 南苑兵营司令部旧址 | 1990.2.23 | 北京市 | 丰台区 | 南苑镇南苑机场 | 清一民国 | 近现代重要史迹及代表性建筑 |

## 房山区

| 总序号 | 保护单位名称 | 公布时间 | 市 | 县(区) | 地点 | 年代 | 类别 |
|---|---|---|---|---|---|---|---|
| 1 | 窦店土城 | 1979.8.21 | 北京市 | 房山区 | 窦店镇 | 战国—西汉 | 古遗址 |
| 2 | 蔡庄土城遗址 | 2001.7.12 | 北京市 | 房山区 | 大石窝镇蔡庄村 | 战国 | 古遗址 |
| 3 | 良乡塔 | 1979.8.21 | 北京市 | 房山区 | 良乡镇东关 | 辽 | 古建筑 |
| 4 | 琉璃河大桥 | 1984.5.24 | 北京市 | 房山区 | 琉璃河镇 | 明 | 古建筑 |
| 5 | 上方山诸寺及云水洞 | 1984.5.24 | 北京市 | 房山区 | 韩村河镇圣水峪村 | 明 | 古建筑 |
| 6 | 姚广孝墓塔 | 1984.5.24 | 北京市 | 房山区 | 青龙湖镇常乐寺村 | 明 | 古建筑 |
| 7 | 玉皇塔 | 1995.10.20 | 北京市 | 房山区 | 大石窝镇高庄村 | 辽 | 古建筑 |
| 8 | 照塔 | 1995.10.20 | 北京市 | 房山区 | 大石窝镇塔照村 | 辽 | 古建筑 |
| 9 | 应公长老寿塔 | 1995.10.20 | 北京市 | 房山区 | 韩村河镇天开村 | 元 | 古建筑 |
| 10 | 周吉祥塔 | 1995.10.20 | 北京市 | 房山区 | 韩村河镇孤山口村 | 明 | 古建筑 |
| 11 | 郊劳台 | 1995.10.20 | 北京市 | 房山区 | 良乡镇大南关 | 清 | 古建筑 |
| 12 | 铁瓦寺 | 2001.7.12 | 北京市 | 房山区 | 河北镇政府 | 明 | 古建筑 |
| 13 | 岫云观 | 2003.12.11 | 北京市 | 房山区 | 琉璃河中学 | 明 | 古建筑 |
| 14 | 白水寺石佛 | 1984.5.24 | 北京市 | 房山区 | 燕山公园 | 元 | 石窟寺及石刻 |
| 15 | 伊桑阿墓石刻 | 2001.7.12 | 北京市 | 房山区 | 韩村河镇皇后台村 | 清 | 石窟寺及石刻 |

## 门头沟区

| 总序号 | 保护单位名称 | 公布时间 | 市 | 县(区) | 地点 | 年代 | 类别 |
|---|---|---|---|---|---|---|---|
| 1 | 沿河城与敌台 | 1984.5.24 | 北京市 | 门头沟区 | 斋堂镇沿河城村 | 明 | 古建筑 |
| 2 | 三官阁过街楼 | 1990.2.23 | 北京市 | 门头沟区 | 龙泉镇琉璃渠村 | 清 | 古建筑 |
| 3 | 灵严寺大殿 | 1995.10.20 | 北京市 | 门头沟区 | 清水镇齐家庄村 | 元 | 古建筑 |
| 4 | 双林寺 | 2001.7.12 | 北京市 | 门头沟区 | 清水镇上清水村西北 | 元—明 | 古建筑 |
| 5 | 天利煤厂旧址 | 2001.7.12 | 北京市 | 门头沟区 | 龙泉镇三家店村中街73、75、77号 | 清 | 古建筑 |
| 6 | 白瀑寺 | 2003.12.11 | 北京市 | 门头沟区 | 雁翅镇淤白村 | 金 | 古建筑 |
| 7 | 灵岳寺 | 2003.12.11 | 北京市 | 门头沟区 | 斋堂镇灵岳寺村 | 清 | 古建筑 |
| 8 | 八路军冀热察挺进军司令部旧址 | 1995.10.20 | 北京市 | 门头沟区 | 斋堂镇马兰村 | 1939年 | 近现代重要史迹及代表性建筑 |
| 9 | 宛平县人民抗日战争为国牺牲烈士纪念碑 | 1995.10.20 | 北京市 | 门头沟区 | 斋堂镇九龙头村 | 1946年 | 近现代重要史迹及代表性建筑 |

## 昌平区

| 总序号 | 保护单位名称 | 公布时间 | 市 | 县(区) | 地点 | 年代 | 类别 |
|---|---|---|---|---|---|---|---|
| 1 | 朝宗桥 | 1984.5.24 | 北京市 | 昌平区 | 沙河镇 | 明 | 古建筑 |
| 2 | 白浮泉遗址——九龙池、都龙王庙 | 1990.2.23 | 北京市 | 昌平区 | 城南街道办事处化庄村东龙山 | 元—明 | 古建筑 |
| 3 | 巩华城 | 1995.10.20 | 北京市 | 昌平区 | 沙河镇 | 明 | 古建筑 |
| 4 | 和平寺 | 1995.10.20 | 北京市 | 昌平区 | 南口镇花塔村北 | 清 | 古建筑 |

## 顺义区

| 总序号 | 保护单位名称 | 公布时间 | 市 | 县(区) | 地点 | 年代 | 类别 |
|---|---|---|---|---|---|---|---|
| 1 | 元圣宫 | 1995.10.20 | 北京市 | 顺义区 | 牛栏山 | 清 | 古建筑 |
| 2 | 无梁阁 | 2001.7.12 | 北京市 | 顺义区 | 大孙各庄镇顾庄子村东 | 清 | 古建筑 |
| 3 | 焦庄户地道战遗址 | 1979.8.21 | 北京市 | 顺义区 | 龙湾屯镇焦庄户村 | 民国 | 近现代重要史迹及代表性建筑 |

## 延庆县

| 总序号 | 保护单位名称 | 公布时间 | 市 | 县(区) | 地点 | 年代 | 类别 |
|---|---|---|---|---|---|---|---|
| 1 | 古崖居遗址 | 1990.2.23 | 北京市 | 延庆县 | 张山营乡 | 唐 | 古遗址 |
| 2 | 玉皇庙山戎墓遗址 | 1995.10.20 | 北京市 | 延庆县 | 张山营乡玉皇庙村 | 东周 | 古墓葬 |
| 3 | 岔道城遗址 | 2001.7.12 | 北京市 | 延庆县 | 八达岭镇岔道村 | 明 | 古建筑 |
| 4 | 永宁天主教堂 | 2001.7.12 | 北京市 | 延庆县 | 永宁镇阜民街 | 清 | 古建筑 |
| 5 | 詹天佑铜像及墓 | 1984.5.24 | 北京市 | 延庆县 | 青龙桥 | 民国 | 近现代重要史迹及代表性建筑 |
| 6 | 木化石群 | 2001.7.12 | 北京市 | 延庆县 | 千家店乡辛栅子村 | 侏罗纪晚期 | 其他 |

## 怀柔区

| 总序号 | 保护单位名称 | 公布时间 | 市 | 县(区) | 地点 | 年代 | 类别 |
|---|---|---|---|---|---|---|---|
| 1 | 红螺寺 | 1990.2.23 | 北京市 | 怀柔区 | 红螺山 | 唐 | 古建筑 |

## 密云县

| 总序号 | 保护单位名称 | 公布时间 | 市 | 县(区) | 地点 | 年代 | 类别 |
|---|---|---|---|---|---|---|---|
| 1 | 白龙潭龙泉寺 | 1995.10.20 | 北京市 | 密云县 | 太师屯镇龙潭沟村 | 清—民国 | 古建筑 |
| 2 | 番字石刻 | 1990.2.23 | 北京市 | 密云县 | 冯家峪镇番字牌村 | 元 | 石窟寺及石刻 |
| 3 | 古北口战役阵亡将士公墓 | 1995.10.20 | 北京市 | 密云县 | 古北口镇南关村 | 民国 | 近现代重要史迹及代表性建筑 |
| 4 | 白乙化烈士陵园 | 1995.10.20 | 北京市 | 密云县 | 石城镇河北村 | 1984年 | 近现代重要史迹及代表性建筑 |

## 平谷区

| 总序号 | 保护单位名称 | 公布时间 | 市 | 县(区) | 地点 | 年代 | 类别 |
|---|---|---|---|---|---|---|---|
| 1 | 上宅文化遗址 | 2001.7.12 | 北京市 | 平谷区 | 金海湖镇上宅村、大兴庄镇北埝头村 | 新石器 | 古遗址 |
| 2 | 丫髻山碧霞元君祠遗址 | 2001.7.12 | 北京市 | 平谷区 | 刘家店镇北吉山村 | 清 | 古遗址 |
| 3 | 鱼子山抗战遗址 | 2001.7.12 | 北京市 | 平谷区 | 山东庄镇鱼子山村 | 1938年 | 近现代重要史迹及代表性建筑 |

## 通州区

| 总序号 | 保护单位名称 | 公布时间 | 市 | 县(区) | 地点 | 年代 | 类别 |
|---|---|---|---|---|---|---|---|
| 1 | 李卓吾墓 | 1984.5.24 | 北京市 | 通州区 | 新华街道办西海子西街12号 | 清 | 古墓葬 |
| 2 | 燃灯塔 | 1979.8.21 | 北京市 | 通州区 | 新华街道办西海子西街12号 | 辽—明 | 古建筑 |

续表：

| 总序号 | 保护单位名称 | 公布时间 | 市 | 县(区) | 地点 | 年代 | 类别 |
|---|---|---|---|---|---|---|---|
| 3 | 通运桥及张家湾镇城墙遗迹 | 1995.10.20 | 北京市 | 通州区 | 张家湾镇张家湾村 | 明 | 古建筑 |
| 4 | 通州清真寺 | 1995.10.20 | 北京市 | 通州区 | 中仓街道办清真寺街1号 | 清 | 古建筑 |
| 5 | 潞河中学原教学楼 | 1990.2.23 | 北京市 | 通州区 | 北苑街道办新城南关31号 | 清、民国 | 近现代重要史迹及代表性建筑 |
| 6 | 富育女校教士楼、百友楼旧址 | 2001.7.12 | 北京市 | 通州区 | 北苑街道玉带河大街72号 | 1904—1929年 | 近现代重要史迹及代表性建筑 |

## 大兴区

| 总序号 | 保护单位名称 | 公布时间 | 市 | 县(区) | 地点 | 年代 | 类别 |
|---|---|---|---|---|---|---|---|
| 1 | 团河行宫遗址 | 2001.7.12 | 北京市 | 大兴区 | 西红门镇团河村 | 清 | 古遗址 |
| 2 | 无碍禅师塔 | 2003.12.11 | 北京市 | 大兴区 | 榆垡乡履磕村 | 金 | 古建筑 |

## 跨区

| 总序号 | 保护单位名称 | 公布时间 | 市 | 地点 | 年代 | 类别 |
|---|---|---|---|---|---|---|
| 1 | 皇城墙遗址 | 2003.12.11 | 北京市 | 东城区、西城区 | 明 | 古遗址 |
| 2 | 明北京城城墙遗迹 | 1984.5.24 | 北京市 | 西城区、东城区 西便门，崇文门东 | 明 | 古建筑 |

# 区县级文物保护单位

## 宣武区

| 总序号 | 保护单位名称 | 公布时间 | 市 | 县(区) | 地点 | 年代 | 类别 |
|---|---|---|---|---|---|---|---|
| 1 | 宝应寺 | 1986.12 | 北京市 | 宣武区 | 登莱胡同29号 | 明 | 古建筑 |
| 2 | 崇效寺藏经阁 | 1990.12 | 北京市 | 宣武区 | 白广路乙27号 | 明 | 古建筑 |
| 3 | 东南园四合院 | 1990.12 | 北京市 | 宣武区 | 东南园胡同49号 | 清 | 古建筑 |
| 4 | 北师大附小旧址 | 1986.12 | 北京市 | 宣武区 | 南新华街20号 | 近现代 | 近现代重要史迹及代表性建筑 |
| 5 | 北师大旧址 | 1986.12 | 北京市 | 宣武区 | 南新华街15号 | 近现代 | 近现代重要史迹及代表性建筑 |
| 6 | 尚小云故居 | 1986.12 | 北京市 | 宣武区 | | 近现代 | 近现代重要史迹及代表性建筑 |
| 7 | 谭嗣同故居 | 1986.12 | 北京市 | 宣武区 | 北半截胡同41号 | 近现代 | 近现代重要史迹及代表性建筑 |
| 8 | 荀慧生故居 | 1986.12 | 北京市 | 宣武区 | 山西街甲13号 | 近现代 | 近现代重要史迹及代表性建筑 |
| 9 | 余叔岩故居 | 1986.12 | 北京市 | 宣武区 | | 近现代 | 近现代重要史迹及代表性建筑 |
| 10 | 粤东新馆 | 1986.12 | 北京市 | 宣武区 | 南横西街11号 | 近现代 | 近现代重要史迹及代表性建筑 |
| 11 | 林白水故居 | 1990.12 | 北京市 | 宣武区 | | 近现代 | 近现代重要史迹及代表性建筑 |
| 12 | 绍兴会馆 | 1990.12 | 北京市 | 宣武区 | 南半截胡同7号 | 近现代 | 近现代重要史迹及代表性建筑 |
| 13 | 沈家本故居 | 1990.12 | 北京市 | 宣武区 | 金井胡同1号 | 近现代 | 近现代重要史迹及代表性建筑 |

## 崇文区

| 总序号 | 保护单位名称 | 公布时间 | 市 | 县(区) | 地点 | 年代 | 类别 |
|---|---|---|---|---|---|---|---|
| 1 | 花市清真寺 | 1984.1.16 | 北京市 | 崇文区 | 西花市大街30号 | 明 | 古建筑 |
| 2 | 夕照寺 | 1984.1.16 | 北京市 | 崇文区 | 夕照寺中街13号 | 明 | 古建筑 |
| 3 | 兴隆街四合院 | 1984.1.16 | 北京市 | 崇文区 | 东兴隆街52号 | 清 | 古建筑 |
| 4 | 安乐禅林 | 1989.7.10 | 北京市 | 崇文区 | 永外安乐林路63号、琉璃井8号 | 明 | 古建筑 |
| 5 | 法华寺 | 1989.7.10 | 北京市 | 崇文区 | 法华寺街65、67号 | 明 | 古建筑 |
| 6 | 药王庙 | 1989.7.10 | 北京市 | 崇文区 | 天坛东晓市101号 | 明 | 古建筑 |
| 7 | 奋章胡同四合院 | 1989.7.10 | 北京市 | 崇文区 | 奋章胡同53号 | 民国 | 古建筑 |
| 8 | 天主教堂 | 1989.7.10 | 北京市 | 崇文区 | 体育馆路永生巷6号 | 民国 | 古建筑 |
| 9 | 三·一八烈士纪念碑 | 1989.7.10 | 北京市 | 崇文区 | 培新街6号 | 1926年 | 近现代重要史迹及代表性建筑 |

## 东城区

| 总序号 | 保护单位名称 | 公布时间 | 市 | 县(区) | 地点 | 年代 | 类别 |
|---|---|---|---|---|---|---|---|
| 1 | 东皇城根南街32号大宅院 | 1984.1.10 | 北京市 | 东城区 | 东皇城根南街32号 | 清 | 古建筑 |
| 2 | 东四八条71号四合院 | 1984.1.10 | 北京市 | 东城区 | 东四八条71号 | 清 | 古建筑 |
| 3 | 惠王府 | 1984.1.10 | 北京市 | 东城区 | 富强胡同3号 | 清 | 古建筑 |
| 4 | 吉安所遗址 | 1984.1.10 | 北京市 | 东城区 | 吉安所右巷10号 | 清 | 古建筑 |
| 5 | 欧美同学会(普胜寺) | 1984.1.10 | 北京市 | 东城区 | 南河沿大街111号 | 清 | 古建筑 |
| 6 | 史家胡同51号四合院 | 1984.1.10 | 北京市 | 东城区 | 史家胡同51号 | 清 | 古建筑 |
| 7 | 史家胡同53号四合院 | 1984.1.10 | 北京市 | 东城区 | 史家胡同53号 | 清 | 古建筑 |
| 8 | 史家胡同55号四合院 | 1984.1.10 | 北京市 | 东城区 | 史家胡同55号 | 清 | 古建筑 |
| 9 | 通教寺 | 1984.1.10 | 北京市 | 东城区 | 针线胡同19号 | 清 | 古建筑 |
| 10 | 顺天府大堂 | 1986.1.21 | 北京市 | 东城区 | 东公街9号 | 明 | 古建筑 |
| 11 | 板厂胡同27号四合院 | 1986.1.21 | 北京市 | 东城区 | 板厂胡同27号 | 清 | 古建筑 |
| 12 | 东四六条55号四合院 | 1986.1.21 | 北京市 | 东城区 | 东四六条55号 | 清 | 古建筑 |
| 13 | 东四四条5号四合院 | 1986.1.21 | 北京市 | 东城区 | 东四四条5号 | 清 | 古建筑 |
| 14 | 富强胡同6、甲6、23号四合院 | 1986.1.21 | 北京市 | 东城区 | 富强胡同6、甲6、23号 | 清 | 古建筑 |
| 15 | 桂公府 | 1986.1.21 | 北京市 | 东城区 | 芳嘉园胡同11号 | 清 | 古建筑 |
| 16 | 黄米胡同5、7、9号四合院 | 1986.1.21 | 北京市 | 东城区 | 黄米胡同5、7、9号 | 清 | 古建筑 |
| 17 | 荣禄故宅 | 1986.1.21 | 北京市 | 东城区 | 菊儿胡同3、5号、寿比胡同6号 | 清 | 古建筑 |
| 18 | 僧格林沁祠堂 | 1986.1.21 | 北京市 | 东城区 | 地安门东大街47号 | 清 | 古建筑 |
| 19 | 什锦花园19号四合院 | 1986.1.21 | 北京市 | 东城区 | 什锦花园19号 | 清 | 古建筑 |
| 20 | 雨儿胡同13号四合院 | 1986.1.21 | 北京市 | 东城区 | 雨儿胡同13号 | 清 | 古建筑 |
| 21 | 清代邮局旧址 | 1996年增补 | 北京市 | 东城区 | 小报房胡同7号 | 清 | 古建筑 |
| 22 | 宝和店碑 | 1986.1.21 | 北京市 | 东城区 | 五塔寺石刻艺术博物馆 | 明 | 石窟寺及石刻 |
| 23 | 德悟和尚行实碑记 | 1986.1.21 | 北京市 | 东城区 | 多福巷胡同44号 | 明 | 石窟寺及石刻 |
| 24 | 皇帝敕谕碑(梓庙) | 1986.1.21 | 北京市 | 东城区 | 帽儿胡同21号 | 明 | 石窟寺及石刻 |
| 25 | 慧照寺修建碑记 | 1986.1.21 | 北京市 | 东城区 | 东四十三条19号 | 明 | 石窟寺及石刻 |
| 26 | 御制护国文昌帝君庙旧碑 | 1986.1.21 | 北京市 | 东城区 | 帽儿胡同21号 | 明 | 石窟寺及石刻 |
| 27 | 皇帝敕谕碑 | 1986.1.21 | 北京市 | 东城区 | 钟楼湾临字9号 | 明 | 石窟寺及石刻 钟鼓楼文物保管所 |
| 28 | 慧仙女校碑 | 1986.1.21 | 北京市 | 东城区 | 五塔寺石刻艺术博物馆 | 清 | 石窟寺及石刻 |
| 29 | 乾隆敕建碑 | 1986.1.21 | 北京市 | 东城区 | 五塔寺石刻艺术博物馆 | 清 | 石窟寺及石刻 |
| 30 | 麻线胡同3号大宅院 | 1984.1.10 | 北京市 | 东城区 | 麻线胡同3号 | 清末 | 近现代重要史迹及代表性建筑 |
| 31 | 北总布胡同2号大宅院 | 1984.1.10 | 北京市 | 东城区 | 北总布胡同2号 | 民国 | 近现代重要史迹及代表性建筑 |
| 32 | 东总布胡同53号旧宅院 | 1984.1.10 | 北京市 | 东城区 | 东总布胡同53号 | 民国 | 近现代重要史迹及代表性建筑 |
| 33 | 段祺瑞宅 | 1984.1.10 | 北京市 | 东城区 | 仓南胡同5号 | 民国 | 近现代重要史迹及代表性建筑 |
| 34 | 马辉堂花园 | 1984.1.10 | 北京市 | 东城区 | 魏家胡同18号 | 民国 | 近现代重要史迹及代表性建筑 |
| 35 | 杨昌济故居 | 1984.1.10 | 北京市 | 东城区 | 豆腐池胡同15号 | 民国 | 近现代重要史迹及代表性建筑 |
| 36 | 朱启钤故宅 | 1984.1.10 | 北京市 | 东城区 | 赵堂子胡同3号 | 民国 | 近现代重要史迹及代表性建筑 |
| 37 | 蔡元培故居 | 1985.10.8增补 | 北京市 | 东城区 | 东堂子胡同75号 | 民国 | 近现代重要史迹及代表性建筑 |
| 38 | 东外清真寺 | 1986.1.21 | 北京市 | 东城区 | 东直门外察慈小区6号 | 清 | 近现代重要史迹及代表性建筑 |
| 39 | 当铺旧址 | 1986.1.21 | 北京市 | 东城区 | 东直门北小街内门楼胡同3、5号 | 民国 | 近现代重要史迹及代表性建筑 |
| 40 | 梁启超故居 | 1986.1.21 | 北京市 | 东城区 | 北沟沿胡同23号 | 民国 | 近现代重要史迹及代表性建筑 |
| 41 | 欧阳予倩故居 | 1986.1.21 | 北京市 | 东城区 | 张自忠路5号 | 民国 | 近现代重要史迹及代表性建筑 |
| 42 | 田汉故居 | 1986.1.21 | 北京市 | 东城区 | 细管胡同9号 | 民国 | 近现代重要史迹及代表性建筑 |

## 西城区

| 总序号 | 保护单位名称 | 公布时间 | 市 | 县(区) | 地点 | 年代 | 类别 |
|---|---|---|---|---|---|---|---|
| 1 | 银锭桥 | 1989.8.1 | 北京市 | 西城区 | 后海北沿东端 | 金 | 古建筑 |
| 2 | 保安寺 | 1989.8.1 | 北京市 | 西城区 | 地安门西大街133、135号 | 元 | 古建筑 |
| 3 | 普寿寺 | 1989.8.1 | 北京市 | 西城区 | 锦什坊街63号 | 元 | 古建筑 |
| 4 | 玉皇阁 | 1989.8.1 | 北京市 | 西城区 | 育强胡同甲22号 | 元 | 古建筑 |
| 5 | 大藏龙华寺 | 1989.8.1 | 北京市 | 西城区 | 后海北沿23号 | 明 | 古建筑 |
| 6 | 德胜桥 | 1989.8.1 | 北京市 | 西城区 | 德胜门内大街 | 明 | 古建筑 |
| 7 | 广福观 | 1989.8.1 | 北京市 | 西城区 | 烟袋斜街37号、大石碑胡同6号 | 明 | 古建筑 |
| 8 | 净业寺 | 1989.8.1 | 北京市 | 西城区 | 德内西顺城街46号 | 明 | 古建筑 |
| 9 | 普济寺 | 1989.8.1 | 北京市 | 西城区 | 西海南沿48号 | 明 | 古建筑 |
| 10 | 三官庙 | 1989.8.1 | 北京市 | 西城区 | 西海北沿29号 | 明 | 古建筑 |
| 11 | 寿明寺 | 1989.8.1 | 北京市 | 西城区 | 鼓楼西大街79号 | 明 | 古建筑 |
| 12 | 双寺 | 1989.8.1 | 北京市 | 西城区 | 双寺胡同11号 | 明 | 古建筑 |
| 13 | 天寿庵 | 1989.8.1 | 北京市 | 西城区 | 龙头井街42号 | 明 | 古建筑 |
| 14 | 万寿兴隆寺 | 1989.8.1 | 北京市 | 西城区 | 北长街39号 | 明 | 古建筑 |
| 15 | 永寿寺 | 1989.8.1 | 北京市 | 西城区 | 三里河前巷1号 | 明 | 古建筑 |
| 16 | 正觉寺 | 1989.8.1 | 北京市 | 西城区 | 正觉胡同甲9号 | 明 | 古建筑 |
| 17 | 醇亲王府(南府) | 1989.8.1 | 北京市 | 西城区 | 鲍家街43号、宗帽胡同甲2号 | 清 | 古建筑 |
| 18 | 翠花街5号四合院 | 1989.8.1 | 北京市 | 西城区 | 翠花街5号 | 清 | 古建筑 |
| 19 | 棍贝子府花园 | 1989.8.1 | 北京市 | 西城区 | 新街口东街31号 | 清 | 古建筑 |
| 20 | 鉴园 | 1989.8.1 | 北京市 | 西城区 | 小凤翔胡同5号 | 清 | 古建筑 |
| 21 | 旌勇祠 | 1989.8.1 | 北京市 | 西城区 | 旌勇里3号 | 清 | 古建筑 |
| 22 | 魁公府 | 1989.8.1 | 北京市 | 西城区 | 宝产胡同甲23、25、27、29号、赵登禹路58、60号、四根柏胡同18号 | 清 | 古建筑 |
| 23 | 清学部 | 1989.8.1 | 北京市 | 西城区 | 教育街1、3号 | 清 | 古建筑 |
| 24 | 摄政王府马号 | 1989.8.1 | 北京市 | 西城区 | 后海北沿43号 | 清 | 古建筑 |
| 25 | 小石桥胡同24号宅园 | 1989.8.1 | 北京市 | 西城区 | 小石桥胡同24号、后马厂胡同17号 | 清 | 古建筑 |
| 26 | 洵贝勒府 | 1989.8.1 | 北京市 | 西城区 | 背阴胡同37号 | 清 | 古建筑 |
| 27 | 仪亲王府 | 1989.8.1 | 北京市 | 西城区 | 府右街137号、西长安街7号 | 清 | 古建筑 |
| 28 | 永佑庙 | 1989.8.1 | 北京市 | 西城区 | 府右街1、3号 | 清 | 古建筑 |
| 29 | 霭公府 | 1989.8.1 | 北京市 | 西城区 | 西绒线胡同51号 | 清 | 古建筑 |
| 30 | 马尾沟教堂 | 1989.8.1 | 北京市 | 西城区 | 车公庄大街6号 | 清 | 近现代重要史迹及代表性建筑 |
| 31 | 陆谟克堂 | 1989.8.1 | 北京市 | 西城区 | 西直门外大街141号 | 民国 | 近现代重要史迹及代表性建筑 |
| 32 | 张自忠故居 | 1989.8.1 | 北京市 | 西城区 | 府右街丙27号 | 民国 | 近现代重要史迹及代表性建筑 |
| 33 | 元大都下水道 | 1989.8.1 | 北京市 | 西城区 | | 元 | 其他 |

## 朝阳区

| 总序号 | 保护单位名称 | 公布时间 | 市 | 县(区) | 地点 | 年代 | 类别 |
|---|---|---|---|---|---|---|---|
| 1 | 常营清真寺 | 1986.6 | 北京市 | 朝阳区 | 常营乡常营村 | 明 | 古建筑 |
| 2 | 南下坡清真寺 | 1986.6 | 北京市 | 朝阳区 | 朝外南下坡 | 清 | 古建筑 |
| 3 | 山东会馆 | 1986.6 | 北京市 | 朝阳区 | 呼家楼南里2号 | 清 | 古建筑 |
| 4 | 肃慎亲王敬敏坟 | 1986.6 | 北京市 | 朝阳区 | 王四营乡道口村 | 清 | 古建筑 |
| 5 | 显谨亲王衍璜坟 | 1986.6 | 北京市 | 朝阳区 | 潘家园街道 | 清 | 古建筑 |

续表：

| 总序号 | 保护单位名称 | 公布时间 | 市 | 县(区) | 地点 | 年代 | 类别 |
|---|---|---|---|---|---|---|---|
| 6 | 张翼祠堂 | 1986.6 | 北京市 | 朝阳区 | 豆各庄乡豆各庄村 | 清 | 古建筑 |
| 7 | 那桐墓 | 1994.7 | 北京市 | 朝阳区 | 双桥 | 清 | 古建筑 |
| 8 | 马骏烈士墓 | 1986.6 | 北京市 | 朝阳区 | 日坛公园西北角 | 现代 | 近现代重要史迹及代表性建筑 |

## 海淀区

| 总序号 | 保护单位名称 | 公布时间 | 市 | 县(区) | 地点 | 年代 | 类别 |
|---|---|---|---|---|---|---|---|
| 1 | 上方寺遗址 | 1999.1.27 | 北京市 | 海淀区 | 苏家坨镇聂各庄凤凰岭 | 辽 | 古遗址 |
| 2 | 瑞云庵明照洞 | 1999.1.27 | 北京市 | 海淀区 | 苏家坨镇聂各庄凤凰岭 | 明 | 古遗址 |
| 3 | 高粱桥 | 1981.2.13 | 北京市 | 海淀区 | 北下关 | 元 | 古建筑 |
| 4 | 白塔庵塔 | 1981.2.13 | 北京市 | 海淀区 | 西三环北路54号中国画研究院 | 明 | 古建筑 |
| 5 | 普照寺 | 1981.2.13 | 北京市 | 海淀区 | 苏家坨镇徐各庄村 | 明 | 古建筑 |
| 6 | 恩慕寺山门 | 1981.2.13 | 北京市 | 海淀区 | 北京大学西门 | 清 | 古建筑 |
| 7 | 恩佑寺山门 | 1981.2.13 | 北京市 | 海淀区 | 北京大学西门 | 清 | 古建筑 |
| 8 | 清代雕楼 | 1981.2.13 | 北京市 | 海淀区 | 香山路正蓝旗 | 清 | 古建筑 |
| 9 | 广源闸及龙王庙 | 1999.1.27 | 北京市 | 海淀区 | 苏州街广源闸 | 元 | 古建筑 |
| 10 | 北坞金山寺及戏楼 | 1999.1.27 | 北京市 | 海淀区 | 四季青乡北坞村 | 明 | 古建筑 |
| 11 | 金仙庵 | 1999.1.27 | 北京市 | 海淀区 | 苏家坨镇北安河鹫峰 | 明 | 古建筑 |
| 12 | 香岩寺 | 1999.1.27 | 北京市 | 海淀区 | 西北旺镇永丰屯 | 明 | 古建筑 |
| 13 | 承泽园 | 1999.1.27 | 北京市 | 海淀区 | 挂甲屯 | 清 | 古建筑 |
| 14 | 海淀镇彩和坊24号四合院 | 1999.1.27 | 北京市 | 海淀区 | 彩和坊24号 | 清 | 古建筑 |
| 15 | 龙泉寺 | 1999.1.27 | 北京市 | 海淀区 | 苏家坨镇聂各庄凤凰岭 | 清 | 古建筑 |
| 16 | 马甸清真寺 | 1999.1.27 | 北京市 | 海淀区 | 马甸南村7号 | 清 | 古建筑 |
| 17 | 香山八旗高等小学 | 1999.1.27 | 北京市 | 海淀区 | 香山南路 | 清 | 古建筑 |
| 18 | 鹫峰山庄遗址(含响塘庙、秀峰寺、地震台) | 1999.1.27 | 北京市 | 海淀区 | 苏家坨镇北安河村西 | 清—民国 | 古建筑 |
| 19 | 法华寺 | 2001.11 | 北京市 | 海淀区 | 魏公村民族大学西门 | 明 | 古建筑 |
| 20 | 西禅寺 | 2001.11 | 北京市 | 海淀区 | 四季青乡小屯村 | 明 | 古建筑 |
| 21 | 周云端塔 | 2001.11 | 北京市 | 海淀区 | 苏家坨镇徐各庄村 | 明 | 古建筑 |
| 22 | 紫竹院行宫 | 2001.11 | 北京市 | 海淀区 | 紫竹院公园内 | 明 | 古建筑 |
| 23 | 立马关帝庙 | 2001.11 | 北京市 | 海淀区 | 蓝靛厂大街1号 | 清 | 古建筑 |
| 24 | 龙王圣母庙 | 2001.11 | 北京市 | 海淀区 | 上庄乡永泰庄 | 清 | 古建筑 |
| 25 | 妙云寺 | 2001.11 | 北京市 | 海淀区 | 玉泉山西路 | 清 | 古建筑 |
| 26 | 升平署 | 2001.11 | 北京市 | 海淀区 | 中央党校南院 | 清 | 古建筑 |
| 27 | 怡贤亲王祠 | 2001.11 | 北京市 | 海淀区 | 温泉镇白家疃 | 清 | 古建筑 |
| 28 | 蓟门烟树碑 | 1981.2.13 | 北京市 | 海淀区 | 学院路西土城 | 清 | 石窟寺及石刻 |
| 29 | 熊希龄墓园 | 1999.1.27 | 北京市 | 海淀区 | 香山煤厂街 | 民国 | 近现代重要史迹及代表性建筑 |
| 30 | 贝家花园 | 2001.11 | 北京市 | 海淀区 | 苏家坨镇北安河村西 | 民国 | 近现代重要史迹及代表性建筑 |
| 31 | 齐白石墓 | 2001.11 | 北京市 | 海淀区 | 魏公村 | 民国 | 近现代重要史迹及代表性建筑 |
| 32 | 孙传芳墓 | 2001.11 | 北京市 | 海淀区 | 北京植物园 | 民国 | 近现代重要史迹及代表性建筑 |

## 石景山区

| 总序号 | 保护单位名称 | 公布时间 | 市 | 县(区) | 地点 | 年代 | 类别 |
|---|---|---|---|---|---|---|---|
| 1 | 万善桥 | 1983.8.27 | 北京市 | 石景山区 | 黑石头双泉寺村东 | 明 | 古建筑 |
| 2 | 龙泉寺 | 1984.3.9 | 北京市 | 石景山区 | 模式口大街北 | 明 | 古建筑 |
| 3 | 石景山古建群 | 1986.10.9 | 北京市 | 石景山区 | 首钢公司厂区内 | 晋—清 | 古建筑 |
| 4 | 皇姑寺 | 1986.10.9 | 北京市 | 石景山区 | 下庄村 | 明 | 古建筑 |
| 5 | 双泉寺 | 1996.9.1 | 北京市 | 石景山区 | 黑石头双泉寺村东 | 金 | 古建筑 |

续表：

| 总序号 | 保护单位名称 | 公布时间 | 市 | 县(区) | 地点 | 年代 | 类别 |
|---|---|---|---|---|---|---|---|
| 6 | 贤良寺塔院 | 1996.9.1 | 北京市 | 石景山区 | 八大处一处长安寺南侧约200米 | 民国 | 古建筑 |
| 7 | 崇兴庵 | 2001.10.29 | 北京市 | 石景山区 | 鲁谷村北鲁谷小学内 | 明 | 古建筑 |
| 8 | 礼王府 | 2001.10.29 | 北京市 | 石景山区 | 福寿岭铁路疗养院内 | 清 | 古建筑 |
| 9 | 雍正御制碑 | 1983.8.27 | 北京市 | 石景山区 | 首钢制氧厂内 | 清 | 石窟寺及石刻 |
| 10 | 福田公墓 | 1996.9.1 | 北京市 | 石景山区 | 福田村东 | 民国 | 近现代重要史迹及代表性建筑 |
| 11 | 隆恩寺冰川擦痕 | 1983.8.27 | 北京市 | 石景山区 | 五里坨南宫村 | 第四纪 | 其他 |
| 12 | 石景山古井 | 1983.8.27 | 北京市 | 石景山区 | 首钢公司厂区石景山上 | 明 | 其他 |
| 13 | 八大处冰川漂砾 | 1986.10.9 | 北京市 | 石景山区 | 八大处公园五至六处之间 | 第四纪 | 其他 |

## 丰台区

| 总序号 | 保护单位名称 | 公布时间 | 市 | 县(区) | 地点 | 年代 | 类别 |
|---|---|---|---|---|---|---|---|
| 1 | 草桥遗迹 | 2003.2 | 北京市 | 丰台区 | 花乡草桥村 | 清 | 古遗址 |
| 2 | 傅子范墓 | 1984.5 | 北京市 | 丰台区 | 南苑乡大红门久敬庄60号皇亭子招待所 | 民国 | 古墓葬 |
| 3 | 火神庙 | 1984.5 | 北京市 | 丰台区 | 长辛店镇大街196号 | 明 | 古建筑 |
| 4 | 密檐塔 | 1984.5 | 北京市 | 丰台区 | 王佐镇瓦窑村 | 明 | 古建筑 |
| 5 | 中顶庙 | 1984.5 | 北京市 | 丰台区 | 南苑乡西铁营村 | 明 | 古建筑 |
| 6 | 和尚塔 | 1984.5 | 北京市 | 丰台区 | 射击场路8号 | 清 | 古建筑 |
| 7 | 娘娘宫 | 1984.5 | 北京市 | 丰台区 | 长辛店镇大街145号 | 清 | 古建筑 |
| 8 | 清真寺 | 1984.5 | 北京市 | 丰台区 | 长辛店镇大街170号 | 清 | 古建筑 |
| 9 | 老爷庙 | 1986.5 | 北京市 | 丰台区 | 长辛店镇大街128号 | 清 | 古建筑 |
| 10 | 福生寺 | 1999.3 | 北京市 | 丰台区 | 长辛店镇张郭庄村 | 明 | 古建筑 |
| 11 | 达园寺 | 1999.3 | 北京市 | 丰台区 | 花乡于家胡同 | 清 | 古建筑 |
| 12 | 王佐镇老爷庙 | 2003.7 | 北京市 | 丰台区 | 王佐镇南宫村 | 清 | 古建筑 |
| 13 | 大王庙 | 2003.10 | 北京市 | 丰台区 | 老庄子乡北天堂村 | 清 | 古建筑 |
| 14 | 张辅张懋墓前石雕 | 1984.5 | 北京市 | 丰台区 | 长辛店镇吕村连山岗 | 明 | 石窟寺及石刻 |
| 15 | 石五供 | 1984.5 | 北京市 | 丰台区 | 王佐镇侯家峪村 | 清初 | 石窟寺及石刻 |
| 16 | 和隆武碑 | 1999.3 | 北京市 | 丰台区 | 卢沟桥乡西南坟村 | 清 | 石窟寺及石刻 |
| 17 | "二七"烈士墓 | 1984.5 | 北京市 | 丰台区 | 长辛店镇桥西花园 | 现代(1966年) | 近现代重要史迹及代表性建筑 |
| 18 | 赵登禹将军墓 | 1984.5 | 北京市 | 丰台区 | 京石公路卢沟桥西道口西侧 | 现代(1980年) | 近现代重要史迹及代表性建筑 |
| 19 | 万佛延寿寺铜观音像(含石碑) | 1984.5 | 北京市 | 丰台区 | 西四环南路64号丰台体育中心内 | 明 | 其他 |
| 20 | 大红门东门房 | 1984.5 | 北京市 | 丰台区 | 大红门东后街152号 | 清 | 其他 |
| 21 | 歙州阳宅 | 1986.5 | 北京市 | 丰台区 | 南苑乡双庙村 | 明 | 其他 |
| 22 | 极乐峰护国宝塔及佛洞群 | 2003.10 | 北京市 | 丰台区 | 王佐镇西庄店后甫营村 | 明 | 其他 |
| 23 | 水志 | 2003.10 | 北京市 | 丰台区 | 卢沟桥乡水文站上游300米处 | 清 | 其他 |

## 房山区

| 总序号 | 保护单位名称 | 公布时间 | 市 | 县(区) | 地点 | 年代 | 类别 |
|---|---|---|---|---|---|---|---|
| 1 | 镇江营遗址 | 1986.9 | 北京市 | 房山区 | 大石窝镇镇江营村 | 商—西周 | 古遗址 |
| 2 | 长沟土城 | 1986.9 | 北京市 | 房山区 | 长沟镇东长沟村 | 汉 | 古遗址 |
| 3 | 大白玉塘 | 1986.9 | 北京市 | 房山区 | 大石窝镇高庄村 | 汉 | 古遗址 |
| 4 | 张坊地道 | 1986.9 | 北京市 | 房山区 | 张坊镇张坊村 | 辽 | 古遗址 |
| 5 | 蟠桃宫遗址 | 1986.9 | 北京市 | 房山区 | 史家营乡柳林水村 | 民国 | 古遗址 |
| 6 | 望诸君墓 | 1986.9 | 北京市 | 房山区 | 良乡镇富庄村 | 战国 | 古墓葬 |

续表:

| 总序号 | 保护单位名称 | 公布时间 | 市 | 县(区) | 地点 | 年代 | 类别 |
|---|---|---|---|---|---|---|---|
| 7 | 贾岛墓 | 1986.9 | 北京市 | 房山区 | 石楼镇二站村 | 唐 | 古墓葬 |
| 8 | 王爷坟 | 1986.9 | 北京市 | 房山区 | 长沟镇西甘池村 | 清 | 古墓葬 |
| 9 | 七级密檐塔 | 1986.9 | 北京市 | 房山区 | 张坊镇下寺村 | 唐 | 古建筑 |
| 10 | 天开塔 | 1986.9 | 北京市 | 房山区 | 韩村河镇天开村 | 辽 | 古建筑 |
| 11 | 五级密檐灵塔 | 1986.9 | 北京市 | 房山区 | 张坊镇张坊村 | 辽 | 古建筑 |
| 12 | 庄公院 | 1986.9 | 北京市 | 房山区 | 周口店镇娄子水村 | 辽—清 | 古建筑 |
| 13 | 严行大德灵塔 | 1986.9 | 北京市 | 房山区 | 长沟镇西甘池村 | 金 | 古建筑 |
| 14 | 于庄塔 | 1986.9 | 北京市 | 房山区 | 窦店镇于庄村 | 金 | 古建筑 |
| 15 | 长春寺 | 1986.9 | 北京市 | 房山区 | 燕山街道敬老院 | 明 | 古建筑 |
| 16 | 常乐寺 | 1986.9 | 北京市 | 房山区 | 青龙湖镇常乐寺村 | 明 | 古建筑 |
| 17 | 豆各庄塔 | 1986.9 | 北京市 | 房山区 | 青龙湖镇豆各庄村 | 明 | 古建筑 |
| 18 | 弘恩寺 | 1986.9 | 北京市 | 房山区 | 窦店镇望楚村 | 明 | 古建筑 |
| 19 | 环秀禅寺 | 1986.9 | 北京市 | 房山区 | 青龙湖镇晓幼营村 | 明 | 古建筑 |
| 20 | 灵鹫禅寺 | 1986.9 | 北京市 | 房山区 | 青龙湖镇北车营村 | 明 | 古建筑 |
| 21 | 吕祖庙 | 1986.9 | 北京市 | 房山区 | 青龙湖镇上万村 | 明 | 古建筑 |
| 22 | 磨碑寺 | 1986.9 | 北京市 | 房山区 | 大石窝镇岩上村 | 明 | 古建筑 |
| 23 | 元武屯娘娘庙 | 1986.9 | 北京市 | 房山区 | 闫村镇元武屯村 | 明 | 古建筑 |
| 24 | 镇江塔 | 1986.9 | 北京市 | 房山区 | 大石窝镇镇江营村 | 明 | 古建筑 |
| 25 | 良乡文庙 | 1986.9 | 北京市 | 房山区 | 良乡镇南街村 | 明—清 | 古建筑 |
| 26 | 凤凰亭 | 1986.9 | 北京市 | 房山区 | 燕山街道办事处 | 清 | 古建筑 |
| 27 | 金成明墓石牌坊 | 1986.9 | 北京市 | 房山区 | 青龙湖镇北刘庄村 | 清 | 古建筑 |
| 28 | 鲁村关帝阁 | 1986.9 | 北京市 | 房山区 | 良乡镇鲁村 | 清 | 古建筑 |
| 29 | 瑞云寺 | 1986.9 | 北京市 | 房山区 | 史家营乡曹家坊村 | 清 | 古建筑 |
| 30 | 庄亲王墓石牌坊 | 1986.9 | 北京市 | 房山区 | 河北镇磁家务村 | 清 | 古建筑 |
| 31 | 观音像 | 1981.9 | 北京市 | 房山区 | 张坊镇北白岱村 | 明 | 石窟寺及石刻 |
| 32 | 大峪沟摩崖造像 | 1986.9 | 北京市 | 房山区 | 张坊镇大峪沟村 | 唐 | 石窟寺及石刻 |
| 33 | 石狮 | 1986.9 | 北京市 | 房山区 | 大石窝镇岩上村 | 元 | 石窟寺及石刻 |
| 34 | 三合庄摩崖造像 | 1986.9 | 北京市 | 房山区 | 张坊镇三合庄村 | 明 | 石窟寺及石刻 |
| 35 | 乾隆御笔碑 | 1986.9 | 北京市 | 房山区 | 长沟镇西长沟村 | 清 | 石窟寺及石刻 |
| 36 | 燕山公园摩崖造像 | 1995.10 | 北京市 | 房山区 | 燕山公园 | 隋 | 石窟寺及石刻 |
| 37 | 郭士红烈士墓 | 1984.9 | 北京市 | 房山区 | 张坊镇张坊村 | 现代 | 近现代重要史迹及代表性建筑 |
| 38 | 郭永生烈士碑 | 1984.9 | 北京市 | 房山区 | 张坊镇片上村 | 现代 | 近现代重要史迹及代表性建筑 |
| 39 | 兵工厂 | 1986.9 | 北京市 | 房山区 | 蒲洼乡蒲洼村 | 现代 | 近现代重要史迹及代表性建筑 |
| 40 | 大安山烈士碑亭 | 1986.9 | 北京市 | 房山区 | 大安山乡大安山村 | 现代 | 近现代重要史迹及代表性建筑 |
| 41 | 河北烈士碑亭 | 1986.9 | 北京市 | 房山区 | 河北镇河北村 | 现代 | 近现代重要史迹及代表性建筑 |
| 42 | 六烈士纪念碑 | 1986.9 | 北京市 | 房山区 | 十渡镇十渡村 | 现代 | 近现代重要史迹及代表性建筑 |
| 43 | 平西抗日烈士碑 | 1986.9 | 北京市 | 房山区 | 十渡镇十渡村 | 现代 | 近现代重要史迹及代表性建筑 |
| 44 | 王仲民烈士墓 | 1986.9 | 北京市 | 房山区 | 蒲洼乡蒲洼村 | 现代 | 近现代重要史迹及代表性建筑 |
| 45 | 赵然墓 | 1986.9 | 北京市 | 房山区 | 十渡镇西庄村 | 现代 | 近现代重要史迹及代表性建筑 |
| 46 | 铜钟 | 1984.9 | 北京市 | 房山区 | 大石窝镇大石窝村 | 明 | 其他 |
| 47 | 陶井 | 1986.9 | 北京市 | 房山区 | 韩村河镇曹章村 | 汉 | 其他 |

## 门头沟区

| 总序号 | 保护单位名称 | 批次或公布时间市 | | 县(区) | 地点 | 年代 | 类别 |
|---|---|---|---|---|---|---|---|
| 1 | 仰山栖隐寺遗址 | 1981.12.12 | 北京市 | 门头沟区 | 妙峰山镇樱桃沟村 | 金—清 | 古遗址 |
| 2 | 椒园寺遗址及"龙虎"二柏 | 1985.12.3 | 北京市 | 门头沟区 | 龙泉镇龙泉务村 | 明 | 古遗址 |

续表：

| 总序号 | 保护单位名称 | 公布时间 | 市 | 县(区) | 地点 | 年代 | 类别 |
|---|---|---|---|---|---|---|---|
| 3 | 长城砖窑遗址 | 1998.11.23 | 北京市 | 门头沟区 | 斋堂镇柏峪村 | 明 | 古遗址 |
| 4 | 耿聚忠墓 | 1981.12.12 | 北京市 | 门头沟区 | 龙泉镇东龙门村 | 清 | 古墓葬 |
| 5 | 西峰寺载洵地宫及享殿 | 1981.12.12 | 北京市 | 门头沟区 | 永定镇岢罗坨村 | 清 | 古墓葬 |
| 6 | 周自齐墓 | 1981.12.12 | 北京市 | 门头沟区 | 龙泉镇城子村 | 民国 | 古墓葬 |
| 7 | 谭鑫培墓 | 1985.12.3 | 北京市 | 门头沟区 | 永定镇栗园庄村 | 民国 | 古墓葬 |
| 8 | 金代壁画墓 | 1996.7.25 | 北京市 | 门头沟区 | 龙泉镇育新学校 | 金 | 古墓葬 |
| 9 | 桃花庵开山祖师塔 | 1981.12.12 | 北京市 | 门头沟区 | 永定镇黑港村 | 明 | 古建筑 |
| 10 | 万佛堂塔 | 1981.12.12 | 北京市 | 门头沟区 | 永定镇万佛堂村 | 明 | 古建筑 |
| 11 | 斋堂东城门 | 1981.12.12 | 北京市 | 门头沟区 | 斋堂镇东斋堂村 | 明 | 古建筑 |
| 12 | 大悲岩观音寺及碑刻 | 1981.12.12 | 北京市 | 门头沟区 | 斋堂镇向阳口村 | 明—清 | 古建筑 |
| 13 | 妙峰山娘娘庙及灵官殿 | 1981.12.12 | 北京市 | 门头沟区 | 妙峰山镇涧沟村 | 清 | 古建筑 |
| 14 | 圈门戏楼 | 1981.12.12 | 北京市 | 门头沟区 | 龙泉镇圈门村 | 清 | 古建筑 |
| 15 | 三家店龙王庙 | 1981.12.12 | 北京市 | 门头沟区 | 龙泉镇三家店村 | 清 | 古建筑 |
| 16 | 庄士敦"乐静山斋"别墅 | 1981.12.12 | 北京市 | 门头沟区 | 妙峰山镇樱桃沟村 | 清末 | 古建筑 |
| 17 | 龙王观音禅林大殿 | 1985.12.3 | 北京市 | 门头沟区 | 斋堂镇马栏村 | 元—明 | 古建筑 |
| 18 | 白云岩石殿堂 | 1985.12.3 | 北京市 | 门头沟区 | 龙泉镇赵家洼村 | 明 | 古建筑 |
| 19 | 朝阳庵 | 1985.12.3 | 北京市 | 门头沟区 | 军庄镇阳坨村 | 明 | 古建筑 |
| 20 | 龙王庙、戏台及柏抱榆、柏抱桑古树 | 1985.12.3 | 北京市 | 门头沟区 | 斋堂镇灵水村 | 明 | 古建筑 |
| 21 | 密檐宝塔 | 1985.12.3 | 北京市 | 门头沟区 | 斋堂镇狼窝港村 | 明 | 古建筑 |
| 22 | 太古化阳洞石塔 | 1985.12.3 | 北京市 | 门头沟区 | 永定镇秋坡村 | 明 | 古建筑 |
| 23 | 万佛堂过街楼 | 1985.12.3 | 北京市 | 门头沟区 | 永定镇万佛堂村 | 明 | 古建筑 |
| 24 | 大寒岭关城 | 1985.12.3 | 北京市 | 门头沟区 | 斋堂镇煤窝村 | 明—清 | 古建筑 |
| 25 | 王平口关城 | 1985.12.3 | 北京市 | 门头沟区 | 永定镇北岭南区王平口村 | 明—清 | 古建筑 |
| 26 | 峰口庵关城 | 1985.12.3 | 北京市 | 门头沟区 | 龙泉镇官厅 | 清 | 古建筑 |
| 27 | 关帝庙 | 1985.12.3 | 北京市 | 门头沟区 | 龙泉镇琉璃渠村 | 清 | 古建筑 |
| 28 | 琉璃厂商宅院 | 1985.12.3 | 北京市 | 门头沟区 | 龙泉镇琉璃渠村 | 清 | 古建筑 |
| 29 | 妙峰山正路万缘 | 1985.12.3 | 北京市 | 门头沟区 | 龙泉镇琉璃渠村 | 清 | 古建筑 |
| 30 | 娘娘庙及戏台 | 1985.12.3 | 北京市 | 门头沟区 | 雁翅镇大村学校 | 清 | 古建筑 |
| 31 | 圈门过街楼 | 1985.12.3 | 北京市 | 门头沟区 | 龙泉镇门头口村 | 清 | 古建筑 |
| 32 | 圈门窑神庙 | 1985.12.3 | 北京市 | 门头沟区 | 龙泉镇圈门村 | 清 | 古建筑 |
| 33 | 天主教堂 | 1985.12.3 | 北京市 | 门头沟区 | 清水镇张家铺村 | 清 | 古建筑 |
| 34 | 刘鸿瑞宅院 | 1985.12.3 | 北京市 | 门头沟区 | 永定镇石门营村 | 民国 | 古建筑 |
| 35 | 下苇甸龙王庙 | 1996.7.25 | 北京市 | 门头沟区 | 妙峰山镇下苇甸村 | 清 | 古建筑 |
| 36 | 白衣庵 | 1998.11.23 | 北京市 | 门头沟区 | 龙泉镇三家店村 | 清 | 古建筑 |
| 37 | 宝峰寺 | 1998.11.23 | 北京市 | 门头沟区 | 斋堂镇西斋堂村 | 清 | 古建筑 |
| 38 | 东街78号院 | 1998.11.23 | 北京市 | 门头沟区 | 龙泉镇三家店村 | 清 | 古建筑 |
| 39 | 二郎庙 | 1998.11.23 | 北京市 | 门头沟区 | 龙泉镇三家店村 | 清 | 古建筑 |
| 40 | 关帝庙 | 1998.11.23 | 北京市 | 门头沟区 | 斋堂镇川底下村 | 清 | 古建筑 |
| 41 | 广亮院 | 1998.11.23 | 北京市 | 门头沟区 | 斋堂镇川底下村 | 清 | 古建筑 |
| 42 | 过街楼 | 1998.11.23 | 北京市 | 门头沟区 | 军庄镇军庄村 | 清 | 古建筑 |
| 43 | 石甬居 | 1998.11.23 | 北京市 | 门头沟区 | 斋堂镇川底下村 | 清 | 古建筑 |
| 44 | 双店院 | 1998.11.23 | 北京市 | 门头沟区 | 斋堂镇川底下村 | 清 | 古建筑 |
| 45 | 中街59号院 | 1998.11.23 | 北京市 | 门头沟区 | 龙泉镇三家店村 | 清 | 古建筑 |
| 46 | 大魏武定三年刻石 | 1981.12.12 | 北京市 | 门头沟区 | 王平镇河北村 | 北朝东魏 | 石窟寺及石刻 |
| 47 | 崇化寺碑刻 | 1981.12.12 | 北京市 | 门头沟区 | 龙泉镇城子村 | 元—明 | 石窟寺及石刻 |
| 48 | 通仙观碑刻 | 1981.12.12 | 北京市 | 门头沟区 | 清水镇燕家台村 | 元—明 | 石窟寺及石刻 |
| 49 | 摩崖造像群 | 1981.12.12 | 北京市 | 门头沟区 | 永定镇石佛村 | 明 | 石窟寺及石刻 |

续表：

| 总序号 | 保护单位名称 | 公布时间 | 市 | 县(区) | 地点 | 年代 | 类别 |
|---|---|---|---|---|---|---|---|
| 50 | 石古岩摩崖石刻 | 1981.12.12 | 北京市 | 门头沟区 | 王平镇西石古崖村 | 明 | 石窟寺及石刻 |
| 51 | 摩崖对联 | 1981.12.12 | 北京市 | 门头沟区 | 妙峰山镇桃园村 | 民国 | 石窟寺及石刻 |
| 52 | 八路军平西司令部第一驻地 | 1981.12.12 | 北京市 | 门头沟区 | 斋堂镇西斋堂村 | 1939年 | 近现代重要史迹及代表性建筑 |
| 53 | 八路军邓宋支队会师地旧址 | 1996.7.25 | 北京市 | 门头沟区 | 清水镇杜家庄村 | 1939年 | 近现代重要史迹及代表性建筑 |
| 54 | 昌宛专署黄安旧址 | 1996.7.25 | 北京市 | 门头沟区 | 清水镇黄安村 | 1939年 | 近现代重要史迹及代表性建筑 |
| 55 | 冀热察军政委员 | 1996.7.25 | 北京市 | 门头沟区 | 清水镇塔河村 | 1940年 | 近现代重要史迹及代表性建筑 |
| 56 | 挺进军司令部塔河旧址 | 1996.7.25 | 北京市 | 门头沟区 | 清水镇塔河村 | 1940年 | 近现代重要史迹及代表性建筑 |
| 57 | 昌宛专署党校黄安旧址 | 1996.7.25 | 北京市 | 门头沟区 | 清水镇黄安村 | 1946年 | 近现代重要史迹及代表性建筑 |
| 58 | 门头沟最早的党支部 | 1998.11.23 | 北京市 | 门头沟区 | 雁翅镇田庄村 | 1932年 | 近现代重要史迹及代表性建筑 |
| 59 | 冀热察区党委 | 1998.11.23 | 北京市 | 门头沟区 | 斋堂镇大三里村 | 1939年 | 近现代重要史迹及代表性建筑 |
| 60 | 挺进军十团团部 | 1998.11.23 | 北京市 | 门头沟区 | 斋堂镇马栏村 | 1939年 | 近现代重要史迹及代表性建筑 |

## 昌平区

| 总序号 | 保护单位名称 | 公布时间 | 市 | 县(区) | 地点 | 年代 | 类别 |
|---|---|---|---|---|---|---|---|
| 1 | 虎峪城 | 2003.7.28 | 北京市 | 昌平区 | 南口镇虎峪村 | 战国 | 古遗址 |
| 2 | 军都故城 | 2003.7.28 | 北京市 | 昌平区 | 马池口镇土城村 | 战国 | 古遗址 |
| 3 | 马刨泉古长城遗址 | 2003.7.28 | 北京市 | 昌平区 | 流村镇马刨泉村 | 战国 | 古遗址 |
| 4 | 东燕州城 | 2003.7.28 | 北京市 | 昌平区 | 兴寿镇西新城村 | 唐 | 古遗址 |
| 5 | 佛岩寺遗址 | 2003.7.28 | 北京市 | 昌平区 | 南口镇湾子村 | 金 | 古遗址 |
| 6 | 大羊山寺庙遗址 | 2003.7.28 | 北京市 | 昌平区 | 兴寿镇大羊山村 | 明 | 古遗址 |
| 7 | 沟沟崖寺庙遗址 | 2003.7.28 | 北京市 | 昌平区 | 十三陵镇德胜口村 | 明 | 古遗址 |
| 8 | 仙人洞 | 2003.7.28 | 北京市 | 昌平区 | 十三陵镇仙人洞村 | 明 | 古遗址 |
| 9 | 小汤山温泉遗址 | 2003.7.28 | 北京市 | 昌平区 | 小汤山镇疗养院 | 清 | 古遗址 |
| 10 | 何营村伯哈智墓 | 2003.7.28 | 北京市 | 昌平区 | 南邵镇何营村 | 明 | 古墓葬 |
| 11 | 李公墓 | 2003.7.28 | 北京市 | 昌平区 | 南口镇南口村 | 明 | 古墓葬 |
| 12 | 孙公墓 | 2003.7.28 | 北京市 | 昌平区 | 南口镇东园村 | 明 | 古墓葬 |
| 13 | 恭亲王墓 | 2003.7.28 | 北京市 | 昌平区 | 崔村镇麻峪村 | 清 | 古墓葬 |
| 14 | 马国柱墓 | 2003.7.28 | 北京市 | 昌平区 | 南口镇东园子村 | 清 | 古墓葬 |
| 15 | 庆禧亲王家族墓地 | 2003.7.28 | 北京市 | 昌平区 | 流村镇白羊城村 | 清 | 古墓葬 |
| 16 | 杨增新墓 | 2003.7.28 | 北京市 | 昌平区 | 沙河镇南一村 | 民国 | 古墓葬 |
| 17 | 半截塔 | 2003.7.28 | 北京市 | 昌平区 | 东小口镇半截塔村 | 辽 | 古建筑 |
| 18 | 白羊城 | 2003.7.28 | 北京市 | 昌平区 | 流村镇白羊城村 | 明 | 古建筑 |
| 19 | 北庄村石塔 | 2003.7.28 | 北京市 | 昌平区 | 长陵镇北庄村 | 明 | 古建筑 |
| 20 | 长峪城 | 2003.7.28 | 北京市 | 昌平区 | 流村镇长峪城村 | 明 | 古建筑 |
| 21 | 木厂村石佛寺 | 2003.7.28 | 北京市 | 昌平区 | 兴寿镇木厂村 | 明 | 古建筑 |
| 22 | 南口城 | 2003.7.28 | 北京市 | 昌平区 | 南口镇南口村 | 明 | 古建筑 |
| 23 | 上关城 | 2003.7.28 | 北京市 | 昌平区 | 南口镇四桥子村 | 明 | 古建筑 |
| 24 | 阿苏卫村关帝庙 | 2003.7.28 | 北京市 | 昌平区 | 小汤山镇阿苏卫村 | 清 | 古建筑 |
| 25 | 长峪城永兴寺 | 2003.7.28 | 北京市 | 昌平区 | 流村镇长峪城村 | 清 | 古建筑 |
| 26 | 慈悲峪村福庆庵 | 2003.7.28 | 北京市 | 昌平区 | 长陵镇慈悲峪村 | 清 | 古建筑 |
| 27 | 海子村九圣庙 | 2003.7.28 | 北京市 | 昌平区 | 兴寿镇海子村 | 清 | 古建筑 |
| 28 | 黄土东村真武庙 | 2003.7.28 | 北京市 | 昌平区 | 回龙观镇黄土东村 | 清 | 古建筑 |
| 29 | 回龙观村玉光寺 | 2003.7.28 | 北京市 | 昌平区 | 回龙观镇回龙观村 | 清 | 古建筑 |
| 30 | 六街城隍庙 | 2003.7.28 | 北京市 | 昌平区 | 昌平邮政局院内 | 清 | 古建筑 |
| 31 | 南口村清真寺 | 2003.7.28 | 北京市 | 昌平区 | 南口镇南口村 | 清 | 古建筑 |
| 32 | 南一村清真寺 | 2003.7.28 | 北京市 | 昌平区 | 沙河镇南一村 | 清 | 古建筑 |
| 33 | 七里渠村观音庵 | 2003.7.28 | 北京市 | 昌平区 | 沙河镇七里渠村 | 清 | 古建筑 |
| 34 | 沙河镇关帝庙 | 2003.7.28 | 北京市 | 昌平区 | 沙河镇医院 | 清 | 古建筑 |

续表：

| 总序号 | 保护单位名称 | 公布时间 | 市 | 县(区) | 地点 | 年代 | 类别 |
|---|---|---|---|---|---|---|---|
| 35 | 上埝头村九圣庙 | 2003.7.28 | 北京市 | 昌平区 | 马池口镇上埝头村 | 清 | 古建筑 |
| 36 | 桃林村东岳庙 | 2003.7.28 | 北京市 | 昌平区 | 兴寿镇桃林村 | 清 | 古建筑 |
| 37 | 土沟村普庆寺 | 2003.7.28 | 北京市 | 昌平区 | 小汤山镇土沟村 | 清 | 古建筑 |
| 38 | 五街清真寺 | 2003.7.28 | 北京市 | 昌平区 | 城北街道五街 | 清 | 古建筑 |
| 39 | 西贯市清真寺 | 2003.7.28 | 北京市 | 昌平区 | 阳坊镇西贯市村 | 清 | 古建筑 |
| 40 | 西沙屯村药王庙 | 2003.7.28 | 北京市 | 昌平区 | 沙河镇西沙屯村 | 清 | 古建筑 |
| 41 | 西新城村双泉村 | 2003.7.28 | 北京市 | 昌平区 | 兴寿镇西新城村 | 清 | 古建筑 |
| 42 | 阳坊镇药王庙 | 2003.7.28 | 北京市 | 昌平区 | 阳坊镇阳坊村 | 清 | 古建筑 |
| 43 | 一中关帝庙 | 2003.7.28 | 北京市 | 昌平区 | 城北街道东关路 | 清 | 古建筑 |
| 44 | 昭陵村关帝庙 | 2003.7.28 | 北京市 | 昌平区 | 长陵镇昭陵村 | 清 | 古建筑 |
| 45 | 南口宝林寺 | 2003.7.28 | 北京市 | 昌平区 | 南口镇兴隆街 | 民国 | 古建筑 |
| 46 | 文物石刻园 | 2003.7.28 | 北京市 | 昌平区 | 昌平公园内 | 辽一民国 | 石窟寺及石刻 |
| 47 | 南庄村经幢 | 2003.7.28 | 北京市 | 昌平区 | 长陵镇南庄村 | 金 | 石窟寺及石刻 |
| 48 | 摩崖造像 | 2003.7.28 | 北京市 | 昌平区 | 南口镇四桥子村 | 元 | 石窟寺及石刻 |
| 49 | 龙潭石刻 | 2003.7.28 | 北京市 | 昌平区 | 南口镇龙潭村 | 明 | 石窟寺及石刻 |
| 50 | 仙枕石刻 | 2003.7.28 | 北京市 | 昌平区 | 南口镇居庸关村 | 明 | 石窟寺及石刻 |
| 51 | 摩崖石刻 | 2003.7.28 | 北京市 | 昌平区 | 阳坊镇防化学院 | 明一民国 | 石窟寺及石刻 |
| 52 | 嵩年墓碑 | 2003.7.28 | 北京市 | 昌平区 | 东小口镇太平庄村 | 清 | 石窟寺及石刻 |
| 53 | 高崖口烈士纪念碑 | 2003.7.28 | 北京市 | 昌平区 | 流村镇狼儿峪村 | 现代 | 石窟寺及石刻 |
| 54 | 十三陵水库纪念碑 | 2003.7.28 | 北京市 | 昌平区 | 十三陵水库大坝东 | 现代 | 石窟寺及石刻 |
| 55 | 西山惨案纪念碑 | 2003.7.28 | 北京市 | 昌平区 | 流村镇溜石港村 | 现代 | 石窟寺及石刻 |
| 56 | 昌平烈士陵园 | 2003.7.28 | 北京市 | 昌平区 | 城北街道西关路 | 现代 | 近现代重要史迹及代表性建筑 |
| 57 | 桃林烈士陵园 | 2003.7.28 | 北京市 | 昌平区 | 兴寿镇桃林村 | 现代 | 近现代重要史迹及代表性建筑 |
| 58 | 周德纯烈士墓 | 2003.7.28 | 北京市 | 昌平区 | 崔村镇东崔村 | 现代 | 近现代重要史迹及代表性建筑 |

## 顺义区

| 总序号 | 保护单位名称 | 公布时间 | 市 | 县(区) | 地点 | 年代 | 类别 |
|---|---|---|---|---|---|---|---|
| 1 | 古城遗址 | 1984.12 | 北京市 | 顺义区 | 后沙峪镇古城村 | 汉 | 古遗址 |
| 2 | 狐奴县遗址 | 1984.12 | 北京市 | 顺义区 | 北小营镇北府村 | 汉 | 古遗址 |
| 3 | 顺义城垣 | 1984.12 | 北京市 | 顺义区 | 仁和镇太平村 | 明 | 古建筑 |
| 4 | 孔庙元碑 | 1984.12 | 北京市 | 顺义区 | 顺义区文物管理所 | 元 | 石窟寺及石刻 |
| 5 | 和硕亲王碑 | 1984.12 | 北京市 | 顺义区 | 李桥镇王家坟村 | 清 | 石窟寺及石刻 |
| 6 | 顺义烈士陵园 | 1984.12 | 北京市 | 顺义区 | 民政局殡仪馆 | 近代 | 近现代重要史迹及代表性建筑 |

## 延庆县

| 总序号 | 保护单位名称 | 公布时间 | 市 | 县(区) | 地点 | 年代 | 类别 |
|---|---|---|---|---|---|---|---|
| 1 | 古夷舆城遗址 | 1984.6.16 | 北京市 | 延庆县 | 旧县古城 | 汉 | 古遗址 |
| 2 | 古缙山县遗址 | 1984.6.16 | 北京市 | 延庆县 | 旧县 | 辽一金 | 古遗址 |
| 3 | 延庆城遗址 | 1984.6.16 | 北京市 | 延庆县 | 延庆镇 | 明 | 古遗址 |
| 4 | 永宁旧城遗址 | 1984.6.16 | 北京市 | 延庆县 | 永宁镇永宁 | 明 | 古遗址 |
| 5 | 菜木沟旧石器遗址 | 1993.2.1 | 北京市 | 延庆县 | 千家店镇菜木沟村 | 旧石器时代 | 古遗址 |
| 6 | 路家河旧石器遗址 | 1993.2.1 | 北京市 | 延庆县 | 张山营镇路家河村 | 旧石器时代 | 古遗址 |
| 7 | 古家窑新石器遗址 | 1993.2.1 | 北京市 | 延庆县 | 千家店镇古家窑村 | 新石器时代 | 古遗址 |
| 8 | 姚家营洞穴遗址 | 1993.2.1 | 北京市 | 延庆县 | 张山营镇姚家营村 | 唐一辽 | 古遗址 |
| 9 | 应梦寺遗址 | 1993.2.1 | 北京市 | 延庆县 | 张山营镇蒋家堡村 | 辽 | 古遗址 |
| 10 | 八仙洞 | 1993.2.1 | 北京市 | 延庆县 | 松山林场 | | 古遗址 |
| 11 | 商周村落遗址 | 1998.12.8 | 北京市 | 延庆县 | 张山营镇胡家营村 | 商一周 | 古遗址 |
| 12 | 香村营商周遗址 | 1998.12.8 | 北京市 | 延庆县 | 沈家营乡香村营村 | 商一周 | 古遗址 |
| 13 | 佛峪口七孔洞遗址 | 1998.12.8 | 北京市 | 延庆县 | 张山营镇佛峪口村 | 唐一辽 | 古遗址 |

续表：

| 总序号 | 保护单位名称 | 公布时间 | 市 | 县(区) | 地点 | 年代 | 类别 |
|---|---|---|---|---|---|---|---|
| 14 | 狐狈沟洞穴遗址 | 1998.12.8 | 北京市 | 延庆县 | 张山营镇水峪村 | 唐—辽 | 古遗址 |
| 15 | 烂角朝阳洞遗址 | 1998.12.8 | 北京市 | 延庆县 | 张山营镇大庄科村 | 唐—辽 | 古遗址 |
| 16 | 烂角焦家洞遗址 | 1998.12.8 | 北京市 | 延庆县 | 张山营镇大庄科村 | 唐—辽 | 古遗址 |
| 17 | 榆林堡古城遗址 | 1998.12.8 | 北京市 | 延庆县 | 康庄镇榆林堡村 | 明 | 古遗址 |
| 18 | 柳沟城遗址 | 2003.12.29 | 北京市 | 延庆县 | 井庄镇柳沟村 | 明 | 古遗址 |
| 19 | 马营城遗址 | 2003.12.29 | 北京市 | 延庆县 | 康庄镇马营村 | 明 | 古遗址 |
| 20 | 南寨坡遗址 | 2003.12.29 | 北京市 | 延庆县 | 大榆树镇兴宝庄村 | 明 | 古遗址 |
| 21 | 山戎墓葬群 | 1984.6.16 | 北京市 | 延庆县 | 永宁镇等罗家台等 | 春秋 | 古墓葬 |
| 22 | 张乾曜墓 | 1984.6.16 | 北京市 | 延庆县 | 延庆镇古家营村 | 唐 | 古墓葬 |
| 23 | 李尚书坟 | 1984.6.16 | 北京市 | 延庆县 | 延庆镇二区北 | 明 | 古墓葬 |
| 24 | 杜家坟 | 1984.6.16 | 北京市 | 延庆县 | 延庆镇上水磨村 | 清 | 古墓葬 |
| 25 | 山南沟胡家坟 | 1993.2.1 | 北京市 | 延庆县 | 刘斌堡乡山南沟村 | 清 | 古墓葬 |
| 26 | 灵照寺 | 1984.6.16 | 北京市 | 延庆县 | 延庆解放街 | 金—清 | 古建筑 |
| 27 | 东红寺龙王庙 | 1984.6.16 | 北京市 | 延庆县 | 康庄镇东红寺 | 明 | 古建筑 |
| 28 | 土边墙 | 1984.6.16 | 北京市 | 延庆县 | 四海镇 | 明 | 古建筑 |
| 29 | 北关龙王庙 | 1984.6.16 | 北京市 | 延庆县 | 延庆镇北关 | 清 | 古建筑 |
| 30 | 东红寺戏楼 | 1984.6.16 | 北京市 | 延庆县 | 康庄镇东红寺 | 清 | 古建筑 |
| 31 | 胡家营戏楼 | 1984.6.16 | 北京市 | 延庆县 | 张山营镇胡家营村 | 清 | 古建筑 |
| 32 | 积善桥 | 1984.6.16 | 北京市 | 延庆县 | 延庆三里河 | 清 | 古建筑 |
| 33 | 莲花山八仙庙 | 1984.6.16 | 北京市 | 延庆县 | 大庄科乡 | 清 | 古建筑 |
| 34 | 西五里营龙王庙 | 1984.6.16 | 北京市 | 延庆县 | 张山营镇西五里营村 | 清 | 古建筑 |
| 35 | 西五里营戏楼 | 1984.6.16 | 北京市 | 延庆县 | 张山营镇西五里营村 | 清 | 古建筑 |
| 36 | 姚家营戏楼 | 1984.6.16 | 北京市 | 延庆县 | 张山营镇姚家营村 | 清 | 古建筑 |
| 37 | 赵庄关帝庙 | 1984.6.16 | 北京市 | 延庆县 | 延庆赵庄 | 清 | 古建筑 |
| 38 | 中羊坊戏楼 | 1984.6.16 | 北京市 | 延庆县 | 张山营镇中羊坊村 | 清 | 古建筑 |
| 39 | 营城子龙王庙 | 1984.6.16 | 北京市 | 延庆县 | 八达岭营城子 | 民国 | 古建筑 |
| 40 | 金刚寺 | 1993.2.1 | 北京市 | 延庆县 | 龙庆峡 | 元 | 古建筑 |
| 41 | 董家沟佛爷庙 | 1993.2.1 | 北京市 | 延庆县 | 大庄科乡董家沟村 | 明 | 古建筑 |
| 42 | 黄龙潭龙王庙 | 1993.2.1 | 北京市 | 延庆县 | 永宁镇上磨村 | 明 | 古建筑 |
| 43 | 孔化营菩萨庙 | 1993.2.1 | 北京市 | 延庆县 | 永宁镇孔化营村 | 明 | 古建筑 |
| 44 | 神仙院 | 1993.2.1 | 北京市 | 延庆县 | 龙庆峡 | 明 | 古建筑 |
| 45 | 双营城 | 1993.2.1 | 北京市 | 延庆县 | 延庆镇双营村 | 明 | 古建筑 |
| 46 | 花盆关帝庙 | 1993.2.1 | 北京市 | 延庆县 | 千家店镇花盆村 | 清 | 古建筑 |
| 47 | 花盆戏楼 | 1993.2.1 | 北京市 | 延庆县 | 千家店镇花盆村 | 清 | 古建筑 |
| 48 | 沙梁子龙王庙 | 1993.2.1 | 北京市 | 延庆县 | 千家店镇沙梁子村 | 清 | 古建筑 |
| 49 | 花楼 | 1995.11.10 | 北京市 | 延庆县 | 四海镇上花楼村 | 明 | 古建筑 |
| 50 | 千家店朝阳寺 | 1998.12.8 | 北京市 | 延庆县 | 千家店镇千家店村 | 清 | 古建筑 |
| 51 | 石佛寺 | 1998.12.8 | 北京市 | 延庆县 | 八达岭水关长城 | 清 | 古建筑 |
| 52 | 大营村烽火台 | 2003.12.29 | 北京市 | 延庆县 | 康庄镇大营村 | 明 | 古建筑 |
| 53 | 东屯真武庙 | 2003.12.29 | 北京市 | 延庆县 | 延庆镇东屯村 | 清 | 古建筑 |
| 54 | 和平街火神庙 | 2003.12.29 | 北京市 | 延庆县 | 永宁镇和平街 | 清 | 古建筑 |
| 55 | 下营崇善寺 | 2003.12.29 | 北京市 | 延庆县 | 张山营镇下营村 | 清 | 古建筑 |
| 56 | 永宁南关龙王庙、关帝庙建筑群 | 2003.12.29 | 北京市 | 延庆县 | 永宁镇南关 | 清 | 古建筑 |
| 57 | 莲花池石狮 | 1984.6.16 | 北京市 | 延庆县 | 延庆镇莲花池 | 唐 | 石窟寺及石刻 |
| 58 | 缙阳寺功德碑 | 1984.6.16 | 北京市 | 延庆县 | 香营小堡 | 辽 | 石窟寺及石刻 |
| 59 | 石佛洞石佛 | 1984.6.16 | 北京市 | 延庆县 | 八达岭镇岔道东沟村 | 辽、金 | 石窟寺及石刻 |
| 60 | 石峡石狮 | 1984.6.16 | 北京市 | 延庆县 | 八达岭镇石峡村 | 元 | 石窟寺及石刻 |
| 61 | 石佛寺石佛群 | 1984.6.16 | 北京市 | 延庆县 | 八达岭镇石佛寺村 | 元、明 | 石窟寺及石刻 |

续表：

| 总序号 | 保护单位名称 | 公布时间 | 市 | 县(区) | 地点 | 年代 | 类别 |
|---|---|---|---|---|---|---|---|
| 62 | 分修边墙题名碑 | 1984.6.16 | 北京市 | 延庆县 | 八达岭镇水关长城 | 明 | 石窟寺及石刻 |
| 63 | 分修长城题名碑 | 1984.6.16 | 北京市 | 延庆县 | 八达岭长博等 | 明 | 石窟寺及石刻 |
| 64 | 李四官庄石狮 | 1984.6.16 | 北京市 | 延庆县 | 延庆镇李四官庄 | 明 | 石窟寺及石刻 |
| 65 | 清水河分界碑 | 1984.6.16 | 北京市 | 延庆县 | 八达岭岔道 | 明 | 石窟寺及石刻 |
| 66 | 延庆西街石狮 | 1984.6.16 | 北京市 | 延庆县 | 延庆镇西街 | 明 | 石窟寺及石刻 |
| 67 | 儒学训导碑 | 1984.6.16 | 北京市 | 延庆县 | 延庆镇 | 明—清 | 石窟寺及石刻 |
| 68 | 五桂头及弹琴峡 | 1984.6.16 | 北京市 | 延庆县 | 八达岭镇三堡村北 | 明—清 | 石窟寺及石刻 |
| 69 | 大浮坨石狮 | 1984.6.16 | 北京市 | 延庆县 | 八达岭镇大浮坨村 | 清 | 石窟寺及石刻 |
| 70 | 狮子营石狮 | 1984.6.16 | 北京市 | 延庆县 | 永宁镇狮子营 | 清 | 石窟寺及石刻 |
| 71 | 望京石及天险 | 1984.6.16 | 北京市 | 延庆县 | 八达岭镇八达岭村 | 清 | 石窟寺及石刻 |
| 72 | 藏文摩崖石刻 | 1993.2.1 | 北京市 | 延庆县 | 八达岭镇特区院内 | 元 | 石窟寺及石刻 |
| 73 | 烧窑峪摩崖造像 | 1993.2.1 | 北京市 | 延庆县 | 旧县镇烧窑峪村 | 明 | 石窟寺及石刻 |
| 74 | 天门关摩崖石刻 | 1993.2.1 | 北京市 | 延庆县 | 四海镇天门关村 | 明 | 石窟寺及石刻 |
| 75 | 白河堡分界碑 | 1993.2.1 | 北京市 | 延庆县 | 香营水库西壁 | 清 | 石窟寺及石刻 |
| 76 | 出入民人恩准碑 | 1993.2.1 | 北京市 | 延庆县 | 四海镇印刷厂内 | 清 | 石窟寺及石刻 |
| 77 | 文昌宫碑 | 1993.2.1 | 北京市 | 延庆县 | 千家店本镇中学 | 清 | 石窟寺及石刻 |
| 78 | 六郎像摩崖造像 | 1998.12.8 | 北京市 | 延庆县 | 八达岭镇青龙桥 | 元 | 石窟寺及石刻 |
| 79 | 五郎像摩崖造像 | 1998.12.8 | 北京市 | 延庆县 | 八达岭镇三堡村北 | 不详 | 石窟寺及石刻 |
| 80 | 果树园烈士纪念碑 | 1966.4.5 | 北京市 | 延庆县 | 井庄乡果树园村 | 现代 | 近现代重要史迹及代表性建筑 |
| 81 | 二毛子坟 | 1984.6.16 | 北京市 | 延庆县 | 延庆镇老白庙 | 1900年 | 近现代重要史迹及代表性建筑 |
| 82 | 平北疗养所遗址 | 1984.6.16 | 北京市 | 延庆县 | 大庄科乡车岭村西 | 1941年 | 近现代重要史迹及代表性建筑 |
| 83 | 西羊坊惨案纪念碑 | 1984.6.16 | 北京市 | 延庆县 | 张山营镇西羊坊村 | 1941年 | 近现代重要史迹及代表性建筑 |
| 84 | 岔道万人坑 | 1984.6.16 | 北京市 | 延庆县 | 八达岭镇岔道 | 1943年 | 近现代重要史迹及代表性建筑 |
| 85 | 烈士陵园 | 1984.6.16 | 北京市 | 延庆县 | 八达岭镇岔道 | 1954年 | 近现代重要史迹及代表性建筑 |
| 86 | 平北司令部遗址 | 1993.1.1 | 北京市 | 延庆县 | 张山营镇海沟村 | 1941年 | 近现代重要史迹及代表性建筑 |
| 87 | 四海革命烈士碑 | 1993.2.1 | 北京市 | 延庆县 | 四海镇四海中学 | 1949年 | 近现代重要史迹及代表性建筑 |
| 88 | 窑湾烈士纪念碑 | 1993.2.1 | 北京市 | 延庆县 | 井庄乡窑湾 | 现代 | 近现代重要史迹及代表性建筑 |
| 89 | 古城口地堡 | 1995.9.23 | 北京市 | 延庆县 | 旧县等古城等村 | 1937年 | 近现代重要史迹及代表性建筑 |
| 90 | 白龙潭纪念碑 | 1995.11.10 | 北京市 | 延庆县 | 大庄科镇白龙潭村 | 1988年 | 近现代重要史迹及代表性建筑 |
| 91 | 平北抗日纪念碑 | 1995.11.10 | 北京市 | 延庆县 | 旧县古城村西 | 1989年 | 近现代重要史迹及代表性建筑 |
| 92 | 巾帼英雄纪念碑 | 1995.11.10 | 北京市 | 延庆县 | 井庄镇柳沟村 | 1990年 | 近现代重要史迹及代表性建筑 |
| 93 | "五桂头"山洞 | 1998.12.8 | 北京市 | 延庆县 | 八达岭石佛寺南 | 清 | 近现代重要史迹及代表性建筑 |
| 94 | 八达岭碉堡 | 1998.12.8 | 北京市 | 延庆县 | 八达岭林场一带 | 1939年 | 近现代重要史迹及代表性建筑 |
| 95 | 千家店革命烈士碑 | 1998.12.8 | 北京市 | 延庆县 | 千家店桥南路边 | 1970年 | 近现代重要史迹及代表性建筑 |
| 96 | 李明英雄纪念碑 | 1998.12.8 | 北京市 | 延庆县 | 旧县白草洼村 | 1984年 | 近现代重要史迹及代表性建筑 |
| 97 | 八达岭瓮城铁炮 | 1985.1.1 | 北京市 | 延庆县 | 八达岭特区 | 明 | 其他 |
| 98 | 大浮坨铁钟 | 1985.1.1 | 北京市 | 延庆县 | 八达岭大浮坨 | 明 | 其他 |
| 99 | 黑龙潭及其览胜碑 | 1985.1.1 | 北京市 | 延庆县 | 八达岭岔道 | 明 | 其他 |
| 100 | 上卢凤营铁钟 | 1985.1.1 | 北京市 | 延庆县 | 张山营镇上卢凤营村 | 清 | 其他 |
| 101 | 白马泉 | 1985.1.1 | 北京市 | 延庆县 | 延庆三里河 | | 其他 |
| 102 | 东红寺铁钟 | 1993.2.1 | 北京市 | 延庆县 | 康庄镇东红寺村 | 明 | 其他 |
| 103 | 滴水壶 | 1993.2.1 | 北京市 | 延庆县 | 千家店镇沙梁子村 | | 其他 |
| 104 | 劈破石 | 1993.2.1 | 北京市 | 延庆县 | 大庄科乡董家沟村 | | 其他 |
| 105 | 珍珠泉 | 1993.2.1 | 北京市 | 延庆县 | 珍珠泉乡珍珠泉村 | | 其他 |
| 106 | 金鱼池 | 1998.12.8 | 北京市 | 延庆县 | 八达岭镇 | | 其他 |

## 怀柔区

| 总序号 | 保护单位名称 | 公布时间 | 市 | 县(区) | 地点 | 年代 | 类别 |
|---|---|---|---|---|---|---|---|
| 1 | 宰相庄肖拜住墓 | 2000.9 | 北京市 | 怀柔区 | 北房镇宰相庄村西北 | 元 | 古遗址 |
| 2 | 甘涧峪古建筑遗址群 | 2000.9 | 北京市 | 怀柔区 | 怀柔镇甘涧峪村东、西、北沟内 | 明 | 古遗址 |
| 3 | 斜音背观音寺 | 2000.9 | 北京市 | 怀柔区 | 雁栖镇长园村南 | 明 | 古遗址 |
| 4 | 火门洞石塔 | 1984.3 | 北京市 | 怀柔区 | 九渡河镇黄花城村北 | 元 | 古建筑 |
| 5 | 怀柔县旧衙署大门 | 1984.3 | 北京市 | 怀柔区 | 怀柔区政府 | 明 | 古建筑 |
| 6 | 鹞子峪堡 | 1984.3 | 北京市 | 怀柔区 | 九渡河镇二道关村西 | 明 | 古建筑 |
| 7 | 凤翔寺 | 1996.6 | 北京市 | 怀柔区 | 杨宋镇仙台村 | 唐 | 古建筑 |
| 8 | 圣泉山观音寺 | 1996.6 | 北京市 | 怀柔区 | 桥梓镇口头村北山 | 明 | 古建筑 |
| 9 | 朝阳洞庙 | 2004.12 | 北京市 | 怀柔区 | 宝山镇碾子村 | 金 | 古建筑 |
| 10 | 甘为霖碑 | 1984.3 | 北京市 | 怀柔区 | 怀柔镇红螺寺 | 明 | 石窟寺及石刻 |
| 11 | 摩崖石刻 | 1984.3 | 北京市 | 怀柔区 | 渤海镇沙峪北沟村北 | 明 | 石窟寺及石刻 |
| 12 | 哈尔布武墓碑 | 1984.3 | 北京市 | 怀柔区 | 桥梓镇西坟村村西 | 清 | 石窟寺及石刻 |
| 13 | 天华洞 | 1984.3 | 北京市 | 怀柔区 | 渤海镇铁矿峪村北 | 清 | 石窟寺及石刻 |
| 14 | 刘仕绥烈士墓 | 1984.3 | 北京市 | 怀柔区 | 怀柔镇大中富乐村西 | 现代 | 近现代重要史迹及代表性建筑 |
| 15 | 怀北镇银杏树 | 1984.3 | 北京市 | 怀柔区 | 怀北镇政府 | 宋 | 其他 |
| 16 | 南冶古松树 | 1984.3 | 北京市 | 怀柔区 | 渤海镇南冶村西 | 元 | 其他 |

## 密云县

| 总序号 | 保护单位名称 | 批次或公布时间 | 市 | 县(区) | 地点 | 年代 | 类别 |
|---|---|---|---|---|---|---|---|
| 1 | 小水峪瓷窑遗址 | 1983.9 | 北京市 | 密云县 | 西田各庄镇小水峪村 | 辽—金 | 古遗址 |
| 2 | 超胜庵遗址 | 1996.9 | 北京市 | 密云县 | 不老屯镇燕落村 | 唐 | 古遗址 |
| 3 | 西庄窠瓷窑遗址 | 1996.9 | 北京市 | 密云县 | 西田各庄镇西庄窠村 | 辽—金 | 古遗址 |
| 4 | 护城古堤遗址 | 1996.9 | 北京市 | 密云县 | 县城西北 | 清 | 古遗址 |
| 5 | 冶仙塔遗址 | 2000.9 | 北京市 | 密云县 | 檀营乡檀营村 | 辽 | 古遗址 |
| 6 | 密云古城遗址 | 2000.9 | 北京市 | 密云县 | 密云镇 | 明 | 古遗址 |
| 7 | 杨令公庙 | 1983.9 | 北京市 | 密云县 | 古北口镇古北口村 | 辽 | 古建筑 |
| 8 | 大成殿 | 1983.9 | 北京市 | 密云县 | 密云县东街四眼井胡同5号 | 元 | 古建筑 |
| 9 | 吕祖庙 | 1983.9 | 北京市 | 密云县 | 古北口镇河西村 | 清 | 古建筑 |
| 10 | 药王庙 | 1983.9 | 北京市 | 密云县 | 古北口镇古北口村 | 清 | 古建筑 |
| 11 | 大公主府 | 1991.12 | 北京市 | 密云县 | 密云县城西 | 清 | 古建筑 |
| 12 | 白道峪内城堡 | 1996.9 | 北京市 | 密云县 | 西田各庄镇白道峪村 | 明 | 古建筑 |
| 13 | 白马关城堡 | 1996.9 | 北京市 | 密云县 | 冯家峪镇白马关村 | 明 | 古建筑 |
| 14 | 关上城堡 | 1996.9 | 北京市 | 密云县 | 大城子镇关上村 | 明 | 古建筑 |
| 15 | 吉家营城堡 | 1996.9 | 北京市 | 密云县 | 新城子镇吉家营村 | 明 | 古建筑 |
| 16 | 姜毛峪城堡 | 1996.9 | 北京市 | 密云县 | 新城子镇塔沟村 | 明 | 古建筑 |
| 17 | 司马台城堡 | 1996.9 | 北京市 | 密云县 | 古北口镇司马台村 | 明 | 古建筑 |
| 18 | 瘟神庙 | 1996.9 | 北京市 | 密云县 | 古北口镇潮关村 | 明 | 古建筑 |
| 19 | 小口城堡 | 1996.9 | 北京市 | 密云县 | 新城子镇小口村 | 明 | 古建筑 |
| 20 | 财神庙 | 1996.9 | 北京市 | 密云县 | 古北口镇古北口村 | 清 | 古建筑 |
| 21 | 清真寺 | 1996.9 | 北京市 | 密云县 | 古北口镇河西村 | 清 | 古建筑 |
| 22 | 大安寺 | 2000.9 | 北京市 | 密云县 | 不老屯镇白土沟村 | 北齐 | 古建筑 |
| 23 | 青洞山三教寺 | 2000.9 | 北京市 | 密云县 | 太师屯镇许庄子村 | 唐 | 古建筑 |
| 24 | 黍谷山西严寺 | 2000.9 | 北京市 | 密云县 | 河南寨镇荆栗园村 | 辽—金 | 古建筑 |
| 25 | 二柏搭枝庙 | 2000.9 | 北京市 | 密云县 | 太师屯镇令公村 | 明 | 古建筑 |
| 26 | 风台顶道观 | 2000.9 | 北京市 | 密云县 | 河南寨镇荆栗园村 | 明 | 古建筑 |
| 27 | 吉祥庵 | 2000.9 | 北京市 | 密云县 | 不老屯镇边庄子村 | 明 | 古建筑 |

续表：

| 总序号 | 保护单位名称 | 公布时间 | 市 | 县(区) | 地点 | 年代 | 类别 |
|---|---|---|---|---|---|---|---|
| 28 | 上峪城堡 | 2000.9 | 北京市 | 密云县 | 冯家峪镇上峪村 | 明 | 古建筑 |
| 29 | 遥桥峪城堡 | 2000.9 | 北京市 | 密云县 | 新城子镇遥桥峪村 | 明 | 古建筑 |
| 30 | 燕山勒功碑 | 1996.9 | 北京市 | 密云县 | 密云县博物馆 | 明 | 石窟寺及石刻 |
| 31 | 三世佛浮雕造像 | 2000.9 | 北京市 | 密云县 | 太师屯镇龙潭沟村 | 元 | 石窟寺及石刻 |
| 32 | 白土沟古崖居 | 2000.9 | 北京市 | 密云县 | 不老屯镇白土沟村 | 不详 | 石窟寺及石刻 |
| 33 | 云峰山摩崖石刻 | 2000.9 | 北京市 | 密云县 | 不老屯镇燕落村 | 不详 | 石窟寺及石刻 |
| 34 | 白乙化烈士牺牲地 | 1983.9 | 北京市 | 密云县 | 石城镇河北村 | 民国 | 近现代重要史迹及将士纪念碑 |
| 35 | 丰滦密联合县政府遗址 | 1983.9 | 北京市 | 密云县 | 西田各庄镇牛盆峪村 | 民国 | 近现代重要史迹及代表性建筑 |
| 36 | 承兴密联合县政府旧址 | 1996.9 | 北京市 | 密云县 | 北庄镇大岭村 | 民国 | 近现代重要史迹及代表性建筑 |
| 37 | 古北口保卫战阵亡将士纪念碑 | 2000.9 | 北京市 | 密云县 | 古北口镇古北口村 | 民国 | 近现代重要史迹及代表性建筑 |
| 38 | 古柏(九搂十八杈) | 1983.9 | 北京市 | 密云县 | 新城子镇新城子村 | 汉 | 其他 |
| 39 | 千年白果树 | 1983.9 | 北京市 | 密云县 | 巨各庄镇久远庄村 | 唐 | 其他 |

## 平谷区

| 总序号 | 保护单位名称 | 公布时间 | 市 | 县(区) | 地点 | 年代 | 类别 |
|---|---|---|---|---|---|---|---|
| 1 | 龙坡遗址 | 1983.6 | 北京市 | 平谷区 | 夏各庄镇安固村 | 商—周 | 古遗址 |
| 2 | 将军关城垣遗址 | 1983.6 | 北京市 | 平谷区 | 金海湖镇将军关村 | 明 | 古遗址 |
| 3 | 杜辛庄遗址 | 1985.6 | 北京市 | 平谷区 | 兴谷街道杜辛庄村 | 商 | 古遗址 |
| 4 | 汉城遗址 | 1985.6 | 北京市 | 平谷区 | 大兴庄镇北城子村 | 汉 | 古遗址 |
| 5 | 峨嵋山营遗址 | 1985.6 | 北京市 | 平谷区 | 南独乐河镇峨嵋山村 | 明 | 古遗址 |
| 6 | 北张岱汉墓 | 1985.6 | 北京市 | 平谷区 | 东高村镇北张岱村 | 汉 | 古墓葬 |
| 7 | 放光汉墓 | 1985.6 | 北京市 | 平谷区 | 王辛庄镇放光村 | 汉 | 古墓葬 |
| 8 | 烽堠 | 1983.6 | 北京市 | 平谷区 | 镇罗营镇上营村 | 明 | 古建筑 |
| 9 | 四座楼 | 1983.6 | 北京市 | 平谷区 | 熊儿寨乡四座楼林场 | 明 | 古建筑 |
| 10 | 文峰塔 | 1983.6 | 北京市 | 平谷区 | 东高村镇东高村 | 清 | 古建筑 |
| 11 | 菩萨庙 | 1985.6 | 北京市 | 平谷区 | 金海湖镇靠山集村 | 明 | 古建筑 |
| 12 | 烈虎桥 | 1994.6 | 北京市 | 平谷区 | 峪口镇东樊各庄 | 明 | 古建筑 |
| 13 | 临泉寺 | 1996.6 | 北京市 | 平谷区 | 东高村镇东高村中心小学 | 清 | 古建筑 |
| 14 | 兴隆庵 | 2002.9 | 北京市 | 平谷区 | 平谷镇北台头村 | 明 | 古建筑 |
| 15 | 平谷石刻艺术馆 | 2002.9 | 北京市 | 平谷区 | 金海湖镇上宅文化陈列馆 | 汉—现代 | 石窟寺及石刻 |
| 16 | 大兴隆禅寺碑刻 | 2002.9 | 北京市 | 平谷区 | 王辛庄镇太后村 | 元 | 石窟寺及石刻 |

## 通州区

| 总序号 | 保护单位名称 | 公布时间 | 市 | 县(区) | 地点 | 年代 | 类别 |
|---|---|---|---|---|---|---|---|
| 1 | 北齐土长城遗址 | 2001.9 | 北京市 | 通州区 | 永顺镇窑厂村 | 北齐 | 古遗址 |
| 2 | 三大冢 | 1959.7 | 北京市 | 通州区 | 永乐店镇德仁务奶牛厂 | 汉 | 古墓葬 |
| 3 | 晾鹰台 | 1985.9 | 北京市 | 通州区 | 永乐店镇德仁务后街村 | 辽 | 古墓葬 |
| 4 | 黄带子坟 | 1985.9 | 北京市 | 通州区 | 马桥镇西马各庄村东北600米 | 清 | 古墓葬 |
| 5 | 潞县东门石桥 | 1985.9 | 北京市 | 通州区 | 潞县镇潞县村东门外 | 清 | 古建筑 |
| 6 | 静安寺 | 1985.9 | 北京市 | 通州区 | 新华街道办静安寺胡同12号 | 清 | 古建筑 |
| 7 | 马驹桥清真寺 | 1985.9 | 北京市 | 通州区 | 马驹桥镇北门口村 | 清 | 古建筑 |
| 8 | 三士庙 | 1985.9 | 北京市 | 通州区 | 张家湾镇陆辛庄村 | 清 | 古建筑 |
| 9 | 于家务清真寺 | 1985.9 | 北京市 | 通州区 | 于家务乡于家务村 | 清 | 古建筑 |
| 10 | 张家湾清真寺 | 1985.9 | 北京市 | 通州区 | 张家湾镇张家湾村 | 清 | 古建筑 |

续表：

| 总序号 | 保护单位名称 | 公布时间 | 市 | 县(区) | 地点 | 年代 | 类别 |
|---|---|---|---|---|---|---|---|
| 11 | 万字会院 | 1985.9 | 北京市 | 通州区 | 中仓街道办西大街9号 | 民国 | 古建筑 |
| 12 | 大成殿 | 2001.9 | 北京市 | 通州区 | 大成街1号 | 清 | 古建筑 |
| 13 | 三义庙 | 2001.9 | 北京市 | 通州区 | 玉带路54号 | 清 | 古建筑 |
| 14 | 通州起义指挥部旧址 | 2001.9 | 北京市 | 通州区 | 东大街83号 | 清 | 古建筑 |
| 15 | 王芝祥故居 | 2001.9 | 北京市 | 通州区 | 新城南街9号 | 清 | 古建筑 |
| 16 | 真武庙 | 2001.9 | 北京市 | 通州区 | 于家务乡北辛店村 | 清 | 古建筑 |
| 17 | 紫清宫 | 2001.9 | 北京市 | 通州区 | 大成街1号 | 清 | 古建筑 |
| 18 | 土桥镇水兽 | 1959.7 | 北京市 | 通州区 | 张家湾镇土桥村 | 明 | 石窟寺及石刻 |
| 19 | 石像生群 | 2001.9 | 北京市 | 通州区 | 西海子西街12号 | 明 | 石窟寺及石刻 |
| 20 | 乾隆御制徐元梦墓碑 | 2001.9 | 北京市 | 通州区 | 宋庄镇草寺村 | 清 | 石窟寺及石刻 |
| 21 | 谕祭国柱碑、文孚碑 | 2001.9 | 北京市 | 通州区 | 梨园镇北杨洼村 | 清 | 石窟寺及石刻 |
| 22 | 日本侵华罪证碑 | 2001.9 | 北京市 | 通州区 | 梨园镇小街村南 | 民国 | 石窟寺及石刻 |
| 23 | 张家门楼 | 1985.9 | 北京市 | 通州区 | 梨园镇北杨洼村 | 民国 | 近现代重要史迹及代表性建筑 |
| 24 | 协和书院教士楼 | 2001.9 | 北京市 | 通州区 | 玉带河大街甲66号 | 清 | 近现代重要史迹及代表性建筑 |
| 25 | 博唐亭 | 2001.9 | 北京市 | 通州区 | 新城南关31号 | 民国 | 近现代重要史迹及代表性建筑 |
| 26 | 冯玉祥驻通营盘 | 2001.9 | 北京市 | 通州区 | 永顺镇窑厂村10号 | 民国 | 近现代重要史迹及代表性建筑 |
| 27 | 平津战役指挥部旧址 | 2001.9 | 北京市 | 通州区 | 宋庄镇宋庄村 | 民国 | 近现代重要史迹及代表性建筑 |
| 28 | 通永道署铁狮 | 1959.7 | 北京市 | 通州区 | 中仓街道办西大街9号 | 元 | 其他 |
| 29 | 宝光寺铜钟 | 1959.7 | 北京市 | 通州区 | 中仓街道办西大街9号 | 明 | 其他 |
| 30 | 马家村古槐 | 1985.9 | 北京市 | 通州区 | 马驹桥镇马村 | 辽 | 其他 |
| 31 | 皇木厂古槐 | 1985.9 | 北京市 | 通州区 | 张家湾镇皇木厂村 | 明 | 其他 |

## 大兴区

| 总序号 | 保护单位名称 | 公布时间 | 市 | 县(区) | 地点 | 年代 | 类别 |
|---|---|---|---|---|---|---|---|
| 1 | 晾鹰台 | 1985.3.2 | 北京市 | 大兴区 | 清云店镇北野厂村 | 元 | 古遗址 |
| 2 | 恭勤夫人谢氏墓 | 1985.3.2 | 北京市 | 大兴区 | 榆垡镇黄各庄村 | 清 | 古墓葬 |
| 3 | 东白塔清真寺 | 1985.3.2 | 北京市 | 大兴区 | 安定镇东白塔村 | 明 | 古建筑 |
| 4 | 薛营清真寺 | 1985.3.2 | 北京市 | 大兴区 | 庞各庄镇薛营村 | 明 | 古建筑 |
| 5 | 礼贤清真寺 | 1989.6.15 | 北京市 | 大兴区 | 礼贤镇礼贤村 | 明 | 古建筑 |
| 6 | 西红门清真寺 | 1989.6.15 | 北京市 | 大兴区 | 西红门镇西红门村 | 明 | 古建筑 |
| 7 | 德寿寺碑 | 1985.3.2 | 北京市 | 大兴区 | 旧宫镇旧宫村 | 清 | 石窟寺及石刻 |
| 8 | 宁佑庙碑 | 1985.3.2 | 北京市 | 大兴区 | 瀛海镇忠兴庄村 | 清 | 石窟寺及石刻 |
| 9 | 双柳树昆仑石 | 1985.3.2 | 北京市 | 大兴区 | 旧宫镇毓顺庄村 | 清 | 石窟寺及石刻 |
| 10 | 永定河神祠碑 | 1989.6.15 | 北京市 | 大兴区 | 庞各庄镇赵村 | 清 | 石窟寺及石刻 |
| 11 | 钟音家族墓碑 | 1989.6.15 | 北京市 | 大兴区 | 庞各庄镇 | 清 | 石窟寺及石刻 |
| 12 | 芦城石狮 | 1989.6.15 | 北京市 | 大兴区 | 黄村镇东芦城村 | 明 | 其他 |

# 北京地区古代建筑修缮工程大事记
## Chronicle of Events Concerning the Renovation
## Works of Ancient Buildings in Beijing Area

1929年，故宫太和殿、钦安殿、交泰殿、咸福宫、储秀宫、符望阁等处进行修缮。

1930年，故宫景仁宫、紫禁城东南、西南角楼等处进行修缮。

1932年，故宫慈宁宫、英华殿、神武门等处进行修缮。

1933年，故宫文华殿、武英殿进行修缮。

1934年，故宫军机处、景福宫、景祺阁进行修缮。皇史宬西配殿进行维修。

1935年，景山五亭进行修缮、油饰彩画工程开工。

1939年，故宫雨花阁进行修缮、油饰彩画工程开工。

1946年，故宫三大殿、慈宁宫进行修缮。

1950年9月，中央人民政府拨款对北海团城城墙进行修缮。

1951年，故宫太和门至保和殿屋面等处进行保养。

1952年，正阳门建筑进行整体修缮。

1953年8月21日，万里长城八达岭段修复竣工。

1953年，天坛斋宫修缮竣工。

1953年，故宫修缮皇极殿、宁寿宫、隆宗门、体仁阁、乐寿堂、养性殿竣工。

1954年，碧云寺中山纪念堂进行修缮，资金19.59万元。

1954年9月，颐和园佛香阁修缮及油饰工程竣工。

1956年5月，古观象台整修竣工。

1957年8月，十三陵长陵祾恩门、祾恩殿、明楼三处避雷设备安装竣工。

1958年9月，颐和园长廊油饰彩画整修工程竣工。

1972年，故宫三大殿、后三宫等宫殿进行下架油饰。

1973年6月，北海五龙亭油饰彩画工程竣工，资金4.8万元。

1975年10月，香山昭庙维修竣工1300平方米，资金8.96万元。

1976年4月，颐和园谐趣园油饰工程竣工，资金10万元。

1978年，颐和园画中游维修工程竣工，资金15万余元。

1979年，古观象台进行维修，资金60万元。白塔寺进行维修，资金35万元。

1980年，德胜门箭楼进行维修，资金30万元。孔庙进行维修，资金35万元。卢沟桥进行维修，资金9万元。

1981年，牛街礼拜寺修缮工程开工。东南角楼落架大修，资金45万元。白云观、西山八大处修缮开工。

1982年，居庸关、大钟寺、定陵进行维修。

1983年至1987年10月，北海小西天、极乐世界殿大修，资金239万元。同年，银山塔林、圆明园遗址整修工程开工，宛平县城维修开工，资金50万元。

1984年，颐和园德和园整修油饰工程完工，资金32万元。

1985年，十三陵昭陵进行维修，资金50万元。同年万寿寺(资金40万元)、先农坛太岁殿(资金150万元)、慕田峪长城(资金69万元)、八达岭长城(资金67万元)修缮工程开工。圆明园内黄花阵遗址修复工程开工。

1986年4月，崇礼住宅东院垂花门与西院定静堂进行维修，于翌年4月竣工。

1986年10月，颐和园苏州街修复工程开工，1990年8月竣工，资金989万元。

1986年12月，卢沟桥主体修复工程开工。

1987年9月，司马台长城修缮工程开工，资金76万元。

1988年4月，北海快雪堂修缮工程开工，1990年4月竣工，资金160万元。

1988年7月，颐和园佛香阁整修工程动工，1989年9月13日竣工，资金146万元。

1988年9月，昭陵祾恩殿工程施工，1989年10月竣工。

1989年4月，先农坛太岁殿修复工程开工，资金60万元。

1991年5月，东四清真寺修缮工程开工，同年10月完工，资金50万元。

1991年8月，颐和园西堤景明楼复建工程开工，资金100万元。

1992年4月，天宁寺塔修缮工程开工，1993年5月竣工，资金50万元。

1992年4月，柏林寺维修工程开工，同年11月完工，资金1050万元。

1992年8月，动物园畅观楼维修工程开工，资金300万元。

1992年9月，银山塔林诸塔进行维修，资金100万元。

1992年12月，房山云居寺复建行宫院、千佛殿，资金100万元。

1993年4月，西拨子长城维修工程启动，资金100万元。

1993年5月，湖广会馆整体建筑维修，资金200万元。

1993年9月9日，上方山七十二庵维修工程开工，资金60万元。

1993年9月，万安公墓陵园建筑大修，面积700平方米，资金200万元。

1993年9月，居庸关长城进行大修，面积1万平方米，资金500万元。

1994年6月，十三陵神路维修工程开工，面积1万平方米，资金200万元。

1995年2月，正乙祠戏楼抢修工程开工，资金300万元。

1995年2月，颐和园澹宁堂复建工程开工，资金1020万元。

1995年4月，历代帝王庙维修工程开工，资金50万元。

1995年7月，嵩祝寺部分重建工程开工，面积1000平方米，资金400万元。

1995年8月，钓鱼台国宾馆城楼复建工程开工，资金80万元。

1996年4月，报国寺山门、天王殿、大雄宝殿维修，资金220万元。

1996年9月，故宫紫禁城城墙、雨花阁及三台地面维修，资金726万元。

1997年3月，颐和园畅观堂进行维修，资金300万元。

1997年4月，先农坛具服殿、观耕台进行修缮，资金97万元。

1997年5月，北海公园镜智宝殿进行复建，面积2000平方米，资金1300万元。

1997年7月，故宫奉先殿、宁寿宫花园进行油饰工程，面积1000平方米，资金85万元。

1997年10月，居庸关长城及城关内外修缮工程，面积4万平方米，资金3312.3万元。

1997年11月，故宫咸福宫、城墙台面维修工程启动，资金850万元。

1997年12月，老舍故居建筑修缮工程启动，面积400平方米，资金81万元。

1998年3月，白塔寺山门、钟鼓楼复建工程启动，资金140万元。

1998年4月，恭王府花园修缮东北角，资金98万元。

1998年4月，天坛南神厨中院进行修缮，资金80万元。

1998年6月，白塔寺东路修缮工程启动，面积658平方米，资金150万元。

1998年6月，故宫启动修复宇墙6942平方米，整修泊岸，面积846平方米，资金1924万元。

1998年8月，故宫端门城楼及东西朝房启动油饰彩画工程，面积2000平方米，资金500万元。

1998年9月，潭柘寺山门修缮工程启动，资金450万元。

1998年11月，碧云寺罗汉堂结构屋面等整体加固工程启动，资金417万元。

1998年12月，明代城墙局部修复50米，资金90万元。

1999年1月，先农坛神厨进行全面修缮，资金464.8万元。

1999年3月，天安门城楼进行局部维修，面积4000平方米，同年7月竣工，资金270万元。

1999年3月，大觉寺南庑等3项建筑抢修，资金195万元。

1999年3月，故宫午门墩台、东华门地面铺墁、太和殿台面西侧地面铺墁工程启动，资金215万元，同年11月完工。

1999年4月，历代帝王庙西配殿、圈墙、影壁等进行维修，面积700平方米，同年9月完工，资金195万元。

1999年4月，颐和园长廊下架油饰，清晏舫进行维修，资金150万元。

1999年4月，香山玉华岫复建工程开工，同年10月竣工，资金300万元。

1999年4月，白云观、吕祖殿、文昌殿等建筑进行维修及翻建，资金375万元。

1999年5月，大觉寺大雄宝殿进行大修，资金95万元。

1999年5月，南堂进行抢险养护工程，面积1260平方米，资金200万元。

1999年5月，故宫奉先殿恢复工字廊、中路建筑油饰等工程启动，资金3295万元。

1999年7月，地坛坛墙进行挑顶大修，资金40万元。

1999年7月，戒台寺牡丹院抢修工程开工，资金218万元。

1999年8月，八大处灵光寺复建玉佛殿，资金60万元。

1999年8月，先农坛东坛墙790延长米进行维修，同年10月完工，资金118万元。

1999年8月，宋庆龄故居南楼进行大修，资金81万元。

1999年10月，北海画舫斋进行维修，面积1437平方米，资金142万元。

2000年4月，和敬公主府寝殿油漆彩画工程启动，资金80万元。

2000年4月，白云观三清殿、戒台等建筑进行修缮，资金299万元。

2000年5月，法源寺大雄宝殿挑顶大修，资金78.61万元。

2000年6月，故宫延春阁复建工程启动，资金816万元。

2000年1月10日，孔庙神厨、井亭、东配房、碑亭、围墙等建筑进行修缮，同年6月30日竣工，资金173.28万元。

2000年4月6日，安徽会馆局部落架大修，同年6月完工，资金126.9万元。

2000年5月27日，古观象台台面进行修缮，资金30万元。

2000年7月30日，天坛神乐署、坛墙进行修缮，资金5801万元。

2000年9月，大慧寺大殿抢修、东西配殿等及环境整治工程开工，同年12月完工，资金100万元。

2000年9月，顺天府学大成殿等建筑复建工程启动，资金450万元。

2000年11月14日，后门桥桥栏板、望柱、泊岸等修复加固工程启动，同年12月完工，资金28.18万元。

2000年11月14日，府学胡同36号院文物修复、油饰装修、院落地面铺装等工程开工，2002年8月30日完工，资金747.37万元。

2000年，圆明园遗址围墙修复保护工程开工，2003年10月完工，资金1417万元。

2001年3月20日，历代帝王庙景德崇圣殿等建筑进行修缮，2002年11月15日完工，资金1910.4万元。

2001年4月9日，戒台寺大悲殿耳房等建筑加固修缮工程启动，2002年完工，资金100万元。

2001年4月10日，五塔寺后罩楼、碑廊等修缮工程开工，2002年7月15日竣工，资金1338.3万元。

2001年4月17日，丰台药王庙进行修复和环境整治，同年完工，资金50万元。

2001年4月20日，钟楼油饰加固和鼓楼地面整修工程开工，2002年5月31日完工，资金551万元。

2001年5月1日，镇岗塔进行修缮，同年7月1日完工，资金50.2万元。

2001年5月12日，德寿堂药店文物修复、油饰装修、院落地面维修等工程启动，2004年10月完工，资金112.85万元。

2001年5月30日，琉璃河大桥及泊岸进行修缮，同年9月2日完工，资金311.3万元。

2001年6月1日，田义墓碑亭修缮工程开工，同年7月1日完工，资金22.2万元。

2001年6月25日，醇亲王府修缮工程开工，2002年完工，资金2025.4万元。

2001年7月11日，皇城墙遗址公园进行维修保护，同年11月完工，资金68.81万元。

2001年8月5日，密云段长城望京楼、古北口城关、鹿皮关进行修缮，2002年11月完工，资金108.47万元。

2001年8月7日，元大都土城遗址进行维修保护，2002年10月15日完工，资金135.4万元。

2001年8月16日，金中都城墙遗址进行维修保护，同年9月完工，资金29万元。

2001年9月1日，德胜门城楼值房修复、现存油饰整修工程开工，2003年8月30日完工，资金190.5万元。

2001年10月，恭王府围墙进行修复保护，2002年10月完工，资金42.7万元。

2001年10月15日，克勤郡王府文物建筑修缮工程开工，2002年4月20日完工，资金216万元。

2001年10月18日，宛平城南城墙及敌楼修复工程开工，2002年6月1日完工，资金400万元。

2001年11月8日，团河行宫翠润轩、小桥、泊岸、御碑亭等抢修工程开工，2002年5月30日完工，资金130万元。

2001年11月10日，无碍禅师塔进行修缮，2002年4月30日完工，资金47.9万元。

2001年11月12日，慈善寺卧佛殿、藏经楼、真武庙修复工程开工，2002年7月完工，资金200万元。

2001年11月15日，承恩寺文物建筑进行修缮，2002年11月15日完工，资金303.16万元。

2001年11月20日，郊劳台进行修缮，2002年5月1日完工，资金35万元。

2001年12月10日，昌平都龙王庙正殿、东西配殿进行修缮，2002年6月10日完工，资金40万元。

2001年12月28日，上房山诸寺中虹桥庵、毗卢殿、舍利院进行修缮，2002年9月28日完工，资金70万元。

2001年，报国寺进行抢修，2002年7月13日完工，资金18.9万元。

2001年，白塔寺中路塔体、雨亭、院落地面等设施进行修复，2002年完工，资金100万元。

2001年，法源寺钟鼓楼、天王殿等中路建筑抢修工程开工，2002年完工，资金100万元。

2001年，周口店遗址进行环境整治，2003年7月31日完工，资金165万元。

2001年，菖蒲河公园皇城墙、太庙墙、普胜寺墙进行修缮，2003年完工，资金60万元。

2002年1月28日，智化寺进行维修保护，2002年9月完工，资金42.4万元。

2002年2月27日，普度寺大殿、山门、庙台进行修缮、清理，同年11月19

日完工，资金630.14万元。

2002年3月1日，关岳庙修缮工程开工，同年7月15日完工，资金460万元。

2002年3月1日，先农坛神厨、宰牲亭、庆成宫、坛墙、院落地面等进行修缮，2004年1月完工，资金1786.89万元。

2002年3月8日，通运桥进行修缮，同年6月28日完工，资金92.1万元。

2002年3月15日，大觉寺北玉兰院等建筑修复地面、管线及环境整治，同年7月完工，资金930.3万元。

2002年3月15日，涛贝勒府文物建筑进行修缮，2002年8月26日完工，资金428万元。

2002年4月1日，和平寺抢修工程开工，同年7月31日完工，资金20万元。

2002年4月1日，大慈延福宫进行修缮，同年7月7日完工，资金80.9万元。

2002年4月1日，岔道城部分城墙进行修缮，同年11月28日完工，资金133万元。

2002年4月10日，天宁寺进行修缮，同年9月6日完工，资金240.1万元。

2002年4月15日，皇城墙长安街段进行修缮，同年7月5日完工，资金46.8万元。

2002年4月24日，红螺寺钟、鼓楼复建工程开工，同年年底竣工，资金48.9万元。

2002年4月30日，明北京城墙遗址保护、环境整理工程启动，资金3685.8万元。

2002年5月15日，地坛坛墙进行修缮，同年8月18日完工，资金511万元。

2002年5月28日，纪晓岚故居抢修工程开工，同年10月15日完工，资金140万元。

2002年5月29日，白塔寺西路搬迁居民、单位及修缮工程启动，2003年6月1日完工，资金3620万元。

2002年5月31日，火德真君庙修缮及地面、管线、消防、环境整治工程启动，资金1000万元。

2002年6月1日，袁崇焕祠文物建筑修复工程开工，同年10月完工，资金470万元。

2002年6月10日，大高玄殿牌楼复建、乾元阁抢修工程开工，2003年9月22日完工，资金138万元。

2002年6月20日，沿河城及敌台进行修缮，同年9月20日完工，资金78.2万元。

2002年6月27日，潭柘寺明王殿、大悲殿抢修工程开工，2003年竣工，资金49.9万元。

2002年6月30日，凝和庙进行抢修，同年10月31日完工，资金177.7万元。

2002年7月1日，丰台娘娘庙文物修复，油饰装修、院落地面等工程开工，2003年6月1日完工，资金327.2万元。

2002年7月5日，国子监街四座牌楼修缮工程开工，同年完工，资金33.4万元。

2002年7月15日，国子监二进院文物修缮及一进院地面修复工程开工，2004年完工，资金770万元。

2002年8月12日，川底下村进行修缮，同年10月22日完工，资金30.1万元。

2002年8月12日，将军关长城修缮工程启动，2003年11月完工，资金194.8万元。

2002年8月25日，北新仓进行抢修，同年11月15日完工，资金163.2万元。

2002年9月1日，巩华城修缮工程启动，同年10月9日完工，资金45万元。

2002年9月13日，东四清真寺中院内文物建筑进行修缮，2004年底完工，资金204.9万元。

2002年9月25日，摩诃庵建筑修缮工程开工，资金159.6万元。

2002年9月30日，景泰陵抢修工程启动，2003年6月15日完工，资金24.9万元。

2002年10月1日，万寿寺西路整体大修、中路万佛阁复建工程开工，2003年6月28日完工，资金2580.4万元。

2002年10月8日，前公用胡同15号进行修缮，2003年5月20日完工，资金118.2万元。

2002年10月8日，十三陵德陵进行修缮，2003年完工，资金1000万元。

2002年10月16日，阳平会馆抢险修缮工程启动，资金108万元。

2002年10月20日，金台书院进行修缮，2003年10月完工，资金153.9万元。

2002年10月21日，升平署戏楼抢修工程开工，2003年10月完工，资金100万元。

2003年1月27日，长椿寺进行修缮，资金1462.9万元。

2003年1月30日，历代帝王庙三期修缮工程开工，资金1342.9万元。

2003年2月10日，天安门城台进行修缮，资金653.4万元。

2003年2月15日，法海寺修复藏经楼、药师殿及现存建筑抢修工程开工，资金537.5万元。

2003年2月28日，密云摇桥峪抢险修缮工程开工，资金15万元。

2003年4月3日，朝阳九天普化宫抢险修缮工程开工，资金20.8万元。

2003年4月15日，隆安寺进行修缮，资金204.4万元。

2003年4月17日，福建汀州会馆北馆进行修缮，资金201.1万元。

2003年4月17日，崇文火神庙全院进行修缮，资金165.9万元。

2003年4月23日，土城海淀段、朝阳段(四段)修缮工程启动，资金139.5万元。

2003年8月25日，醇亲王墓(七王坟)抢险修缮工程开工，资金40.8万元。

2003年8月26日，丰台老爷庙抢险修缮工程开工，资金60万元。

2003年9月2日，通州清真寺、通州文庙等修缮工程启动，资金1033.1万元。

2003年9月18日，团团演武厅东朝房、围墙地面进行修复，同年10月15日完工，资金483.2万元。

2003年，十三陵康陵、庆陵、泰陵修缮工程启动，资金330万元。

2003年，长城黄花城、古北口、岔道城、箭扣、吉家营、司马台等处修缮工程启动，资金2557.2万元。

2003年，白塔寺塔身抢险、中路地面铺装等工程启动，资金138.9万元。

2003年，琉璃厂宣武实验小学旧址修缮工程开工，资金100万元。

2003年，金陵遗址神道保护、石棺搬运补偿费等，资金113万元。

2003年，大兴火神庙抢险修缮工程开工，资金35.5万元。

2003年，平谷兴隆庵、菩萨庙抢险修缮工程开工，资金10万元。

2003年，西城三官庙抢险修缮工程开工，资金150万元。

2003年，通县燃灯塔抢险修缮工程开工，资金10万元。

2004年1月，牛街礼拜寺进行修缮，资金621.9万元。

2004年3月9日，宛平城城墙二期修缮、城门抢险修缮等工程开工，资金241.5万元。

2004年3月30日，通州于家务清真寺抢险修缮工程开工，资金100万元。

2004年4月12日，大钟寺地面、彩画修缮工程启动，资金705.4万元。

2004年4月21日，房山照塔、应公长老寿塔等进行修缮，同年10月9日完工，资金29万元。

2004年5月，拈花寺抢险修缮工程开工，资金6.5万元。

2004年5月26日，智化寺转轮藏殿保护工程开工，资金43.5万元。

2004年6月14日，岫云观修缮工程开工，同年11月23日完工，资金100万元。

2004年6月30日，白瀑寺圆正法师塔修缮工程开工，资金33.4万元。

2004年7月28日，燕墩抢险修缮工程开工，资金20万元。

2004年8月，法源寺修缮工程启动，资金671.6万元。

2004年8月，东四九条小学方亭修缮工程开工，资金11.5万元。

2004年9月，月坛抢险修缮工程开工，资金943.2万元。

2004年9月3日，府学胡同36号局部进行修缮，资金52万元。

2004年9月8日，恒亲王府修缮工程开工，资金400万元。

2004年9月13日，丫髻山修缮工程开工，资金950.4万元。

2004年9月15日，东城区天祥祠抢险修缮工程开工，资金10万元。

2004年10月13日，地坛文物修缮，补助350万元。

2004年11月30日，宣仁庙修缮，资金402.5万元。

2004年11月30日，刘秉权墓清理、抢险修缮工程开工，同年12月17日完工，资金15万元。

2004年12月20日，孚郡王墓修缮工程开工，资金314.6万元。

2004年，朱彝尊故居修缮工程开工，资金441万元。

2005年3月14日，延庆古崖居抢险加固工程开工，资金286.9万元。

2005年3月15日，都龙王庙钟鼓楼抢险修缮工程开工，资金22.7万元。

2005年3月31日，上方山兜率寺修缮工程开工，同年11月15日完工，资金87.4万元。

2005年4月5日，八大处大悲寺等三处抢险修缮工程开工，资金100万元。

2005年4月29日，报国寺文物建筑油饰修复工程开工，资金50万元。

2005年5月，凤凰岭玲珑塔抢险修缮工程开工，同年6月完工，资金8.3万元。

2005年6月2日，前鼓楼苑四合院抢险修缮工程开工，资金50万元。

2005年7月，顺天府大堂修缮工程，资金30万元。

2005年7月8日，团城演武厅实胜寺碑亭修缮工程开工，资金103.8万元。

2005年7月12日，颐和园佛香阁、天坛、北海修缮工程启动，资金2250万元。

2005年7月19日，北海琼岛文物建筑修缮工程开工，资金1250万元。

2005年8月，东四清真寺三期修缮工程开工，资金526.8万元。

2005年8月3日，灵岳寺一期、二期修缮工程启动，资金196.8万元。

2005年8月3日，皇城墙抢险修缮工程开工，资金41.7万元。

2005年8月31日，西堂子胡同35号院修缮工程开工，资金511.5万元。

2005年9月14日，北海西天梵境院修缮、市政排污工程开工，资金484.6万元。

2005年9月28日，正阳门及箭楼修缮工程开工，资金948.2万元。

2005年10月9日，万寿寺后罩楼、门前御路修缮工程开工，资金182.1万元。

2005年10月17日，德胜门值房抢险修缮工程开工，资金49.2万元。

2005年10月24日，模式口地区慈善寺、承恩寺、法海寺二期抢险和壁画修缮工程开工，资金1556.7万元。

2005年11月30日，都城隍庙抢险修缮工程，补助60万元。

2005年，国子监第一、三进院古建修缮、彩画修复，资金2436万元。

2005年，天坛祈年殿建筑群修缮，资金1305万元。

2005年，吉安所四合院抢险修缮工程开工，资金53.4万元。

2005年，古观象台抢险修缮工程开工，资金60.2万元。

2005年，密云大公主府抢险修缮工程开工，资金50万元。

2005年，朱圭墓抢险修缮工程开工，资金10万元。

2005年，开元寺抢险修缮工程开工，资金101.9万元。

2005年，西城火神庙一期修缮、后罩楼修缮工程启动，资金1000万元。

2006年3月，戒台寺千佛阁修复、滑坡抢险加固修缮等工程开工，资金800万元。

2006年3月8日，宣武区湖广会馆等整治项目启动，资金140万元。

2006年4月6日，延庆火焰山长城营盘遗址修缮工程开工，资金90万元。

2006年4月27日，雪池冰窖抢险修缮工程开工，资金11万元。

2006年4月28日，朝阳区龙王庙、乌雅氏家族墓碑修缮工程开工，资金101.9万元。

2006年5月30日，太庙修缮工程开工，资金1411.6万元。

2006年6月14日，阳平会馆戏楼二期文物修缮工程开工，资金498.6万元。

2006年6月16日，朝阳区北顶娘娘庙遗址保护古建修缮工程开工，资金667.3万元。

2006年6月23日，上庄东岳庙文物抢险修缮工程开工，资金153.7万元。

2006年6月28日，八达岭残长城修缮工程开工，资金400万元。

2006年7月6日，大觉寺佛像、玉兰院文物及建筑修缮工程开工，资金611.6万元。

2006年8月2日，伊桑阿墓抢险修缮工程开工，资金29.6万元。

2006年8月7日，中南海围墙抢险修缮工程开工，资金540.8万元。

2006年10月10日，宛平城西城墙抢险修缮工程开工，资金93.6万元。

2006年11月21日，团城演武厅油饰、围墙及大门抢险修缮工程开工，资金89.9万元。

2006年，孔庙建筑修缮、彩画修复工程启动，资金2236.6万元。

2006年，醇亲王府(南府)文物修缮工程开工，资金450万元。

2006年，礼贤清真寺抢险修缮工程开工，资金30万元。

2006年，东白塔清真寺抢险修缮工程开工，资金30万元。

2006年，通州静安寺抢险修缮工程开工，资金30万元。

# 后记

  《北京古代建筑精粹》(上、下)由北京市文物局和北京出版社出版集团合作出版，北京市古代建筑研究所负责具体的编写工作。在本书的编写过程中，北京市文物局的各位领导一直非常关心和重视，多次组织召开研讨会，邀请专家学者讨论确定本书的体例及写作大纲，从最初确定的800多处北京古代建筑中择其精品，最终确定102处最有代表性的古代建筑，以飨读者。

  北京是著名的历史文化名城，具有3000多年的建城史和800多年的建都史，作为中国封建王朝最后四个王朝的都城，北京保留至今的古代建筑在全国乃至世界上都是首屈一指的，尤其是保留了全国最多，最完整的明、清建筑群，其中被列为"世界文化遗产"的古代建筑群就有五处。这些古代建筑具有极高的历史人文价值和艺术价值，要加以充分地展现，无形中为编写工作增加了难度和压力。经过两年多的酝酿和艰苦的收集整理资料工作，进行了紧张有序的编写工作，终于付梓。

  本书在编写过程中，得到众多名摄影家的支持，为本书提供了大量高质量的图片资料，很多摄影家还参与了补拍工作，众多古代建筑使用单位积极配合，为补拍工作提供了方便。书中部分线图由张景阳、范磊、刘艳、孙海红、马羽杨、曹志国、刘佳、周颖、忻琳、单杰等绘制和整理。本书还引用了《宣南鸿雪图志》、《北京中轴线建筑实测图典》、《中国古代建筑史》等书中的少量图纸。本书是从事北京古代建筑研究和保护的诸位同人集体心血的结晶，在此，谨向他们表示谢忱，并向所有帮助和关心本书出版的各位朋友致以真诚的谢意。借此机会，还要向为保护北京古代建筑奋斗一生的老一辈古建工作者致意。

  由于我们水平有限，错误之处，在所难免，恳请专家和广大读者给予指正。

责任编辑　程阳阳　　吕　晓　　张承志
文字编辑　钱　颖
图片编辑　张肇基　　张承志　　徐丹瑜
特约编辑　梁玉贵　　李卫伟
文字撰写　侯兆年　　梁玉贵　　刘　珊
　　　　　付　莉　　李卫伟　　刘文丰
　　　　　高　梅
图片摄影　张肇基　　蒙　紫　　李　江
　　　　　杨　茵　　董瑞成　　陈东林
　　　　　楼庆西　　胡敦志　　姜景余
　　　　　高明义　　何炳富　　张承志
　　　　　王玉伟　　徐丹瑜　　王建华
　　　　　吴健骅　　信岩君　　张晨声
　　　　　张振光　　梁再生　　何　伟
　　　　　刘培恩　　张弘强　　王英武
　　　　　马炳坚　　胡维标　　杜殿文
　　　　　许延增　　邢延生　　常胜凯
　　　　　柳　魁　　董文建　　刘文丰
　　　　　梁玉贵　　段晓陕　　黄　鑫
　　　　　李　志　　文　馨　　武裁军
　　　　　贺建淼　　王慧明　　姚天新
线图绘制　张景阳　　范　磊　　刘　艳
　　　　　孙海红　　马羽杨　　曹志国
　　　　　刘　佳　　周　颖　　忻　琳
　　　　　单　杰
版式设计　刘金川　　龚　同
英文翻译　于力凡　　刘　勇
责任印制　李文宗　　毛宇楠